教育部中等职业教育“十二五”国家规划立项教材
中等职业教育服装设计与工艺专业系列教材

服装英语

FUZHUANG YINGYU

主　编/ 黎喜欢　刘　娟
副主编/ 姚景明　郭碧坤　杨柳梅　李秋雯　廖满茹
参　编/ 赵艺宁　骆鋆熳　方　茵　梁永旋　廖赞峰

重庆大学出版社

图书在版编目(CIP)数据

服装英语/黎喜欢，刘娟主编. -- 重庆：重庆大学出版社，2020.9
中等职业教育服装设计与工艺专业系列教材
ISBN 978-7-5689-0708-8

Ⅰ. ①服… Ⅱ. ①黎…②刘… Ⅲ. ①服装工业—英语—中等专业学校—教材 Ⅳ. ①TS941

中国版本图书馆CIP数据核字（2019）第147514号

服装英语

主　编　黎喜欢　刘　娟
副主编　姚景明　郭碧坤　杨柳梅　李秋雯　廖满茹
责任编辑：陈一柳　版式设计：尹　恒
责任校对：刘志刚　责任印制：赵　晟

重庆大学出版社出版发行
出版人：饶帮华
社　址：重庆市沙坪坝区大学城西路21号
邮　编：401331
电　话：（023）88617190　88617185（中小学）
传　真：（023）88617186　88617166
网　址：http://www.cqup.com.cn
邮　箱：fxk@cqup.com.cn（营销中心）
全国新华书店经销
重庆升光电力印务有限公司印刷

开本：787mm×1092mm　1/16　印张：9.75　字数：286千
2020年9月第1版　2020年9月第1次印刷
ISBN 978-7-5689-0708-8　定价：49.00元

本书如有印刷、装订等质量问题，本社负责调换

前　言

随着中国服装贸易外向度的不断深入，强化服装专业英语基础及其应用能力成为服装专业英语教学的重要任务。为此，针对中等职业技术学校服装专业的英语教学，结合服装行业实际工作流程和工作场景，我们开发了这本适合中等职业教育服装设计与工艺专业的服装英语教材。

在编写中，我们选取了中等职业教育服装专业学生就业最常用的职业场景，内容以听说技能训练为主，注重语言在工作场所中的实际运用，设计图文并茂，形式活泼多样，力求把课堂变成学生情感交流、潜能开发、思想沟通、个性发展的大舞台，全方位、多角度地进行职场演练，真正体现“以行业需求为导向，以能力培养为本位，以学习者为中心”的职业教育原则。

本教材共12个单元，每个单元由Warming Up, Listening and Speaking, Reading and Writing, Further Study和My Progress Check组成，设置了不同的职业场景，让学生进行听说读写多项技能的训练。Further Study主要介绍服饰文化，旨在拓宽学生的文化视野，提高审美情趣、文化品位。教材主要培养学生道德情感、文化素养、思维能力、语言能力、交际能力等英语学科核心素养与能力。

本教材的主编是广州市增城区职业技术学校的黎喜欢和中山沙溪理工学校的刘娟两位老师，编者有广州市增城区职业技术学校的姚景明、郭碧坤、杨柳梅、李秋雯、廖满茹、赵艺宁，广州市海珠商务职业学校的骆鋆墁，中山沙溪理工学校的方茵、梁永旋，中山市建勋中学的廖赞峰等。

在编写过程中，我们还得到了各中职学校的领导和老师的大力支持和协助，尤其是广州市增城区职业技术学校教研处郭杰敏主任、服装设计与工艺专业带头人汤杰兰老师等，在此一并表示感谢！

编　者

2020年2月

目　录

UNIT 1

In the Store

【Goals】

- Identify different types of clothes
- Identify the information on the sales list
- Talk about clothes between salespeople and customers
- Identify different parts of the garment
- Identify Chinese traditional garment
- Talk about Chinese traditional garment

Warming Up

1. Look and match.

a. vest	b. blouse	c. overall
d. leggings	e. culottes	f. tuxedo
g. evening dress	h. A-lined skirt	i. princess dress

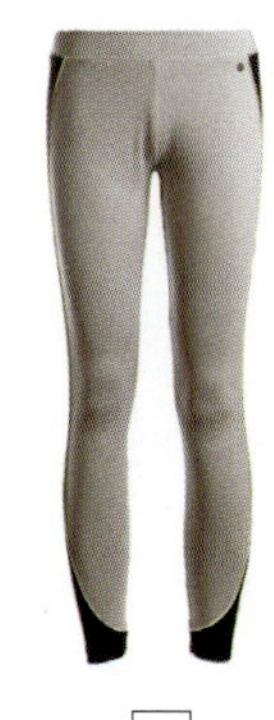

2. Look and translate.

category____________________

standard____________________

style____________________

size____________________

color____________________

grade____________________

country of origin____________________

inspector____________________

safety sort____________________

composition____________________

price____________________

3. Look and tick.

Tick what information you can get on the sales list as much as you can.

Sales List							
Shop: No. 1							Jan, 2017
Date	Item number	Color	Size	Original price (RMB)	Discount	Real price (RMB)	Amount (suit)
1st	B001	Gray	L	2100	10%	1890	6
1st	A001	Black	XL	3000	15%	2550	3

☐ date ☐ customer ☐ discount ☐ item number

☐ size ☐ color ☐ amount ☐ original price

Listening and Speaking

Dialogue One

1. Learn the words.

2. Listen to the dialogue and repeat.

Salesperson: Hello, welcome to David Men's Store. How can I help you?
Tom: I'm looking for a new suit.
Salesperson: Do you know what size you need?
Tom: No, I'm not sure. I'd like to be measured please!
Salesperson: OK. It looks like you need about a 48 size long jacket, and your waist is 36. Which colors are you interested in?
Tom: I'd like to see one in black and another in brown. Thanks a lot.

3. Listen again and tick the answers.

What did the customer say in the store?

☐ How can I help you?
☐ I'm looking for a new jacket.
☐ Do you know what size you need?
☐ I'd like to be measured please!
☐ I'd like to see one in black and another in brown.
☐ It looks like you need about a 48 long jacket.

4. Pair work.

Follow the above example and make a dialogue.

Tips

① I'm looking for a new jacket. 我想买一件夹克衫。
be looking for 表示正在找……，购物时常常会使用，意为“想买……”。

② I'd like to be measured please! 我想先要测量一下。
I'd like to= I would like to 表示我想要……，购物时也常常会使用。
would like to+do，如：I'd like to buy a dress.

Dialogue Two

1. Learn the words.

belt outlet casual dressy leather lighter darker style

2. Listen and decide whether the sentences are true (T) or false (F).

( ) ①Sue wants to buy a bag too.
( ) ②Sue likes a belt that is kind of casual and dressy.

(　　) ③Sue prefers a darker one.

(　　) ④Sue tells the salesperson she likes the one made of leather.

3. Listen again and fill in the blanks.

Salesperson: Hi, welcome to the ________. Can I help you today?

Sue: Sure, I need a belt that is kind of ________ and kind of ________.

Salesperson: Ur, perhaps a ________ belt would work. Do you prefer ________ or darker colors?

Sue: I like darker colors, maybe brown.

Salesperson: OK, here are a few different ________. I'll let you look at them. If you need anything else, just let me know.

Sue: All right, thanks a lot.

Tips

prefer: 更喜欢

Do you prefer lighter or darker colors? 你喜欢深色还是浅色?

常用结构有: prefer A to B 喜欢A多过B

如: I prefer darker to lighter colors. 我喜欢深色多过浅色。

4. Complete the dialogue.

Salesperson: Hello, welcome to Queen store. ____________________, madam? (早上好,女士,我能为你做些什么?)

Sue: I ____________________ (我想买条漂亮的裙子).

Salesperson: ____________________ (你需要什么类型的裙子呢)?

Sue: A dress that is kind of casual.

Salesperson: I see. ____________________ (你对什么颜色感兴趣呢)?

Sue: I prefer lighter color please. Thanks a lot.

5. Role play and talk with your partners with the following sentence patterns.

What can I do for you?
I am looking for...
What size/color do you need?

I prefer...
I'd like to be measured please!
Here are a few different styles.

Reading and Writing

1. Look and discuss.

Can you fill the blanks with the key words?

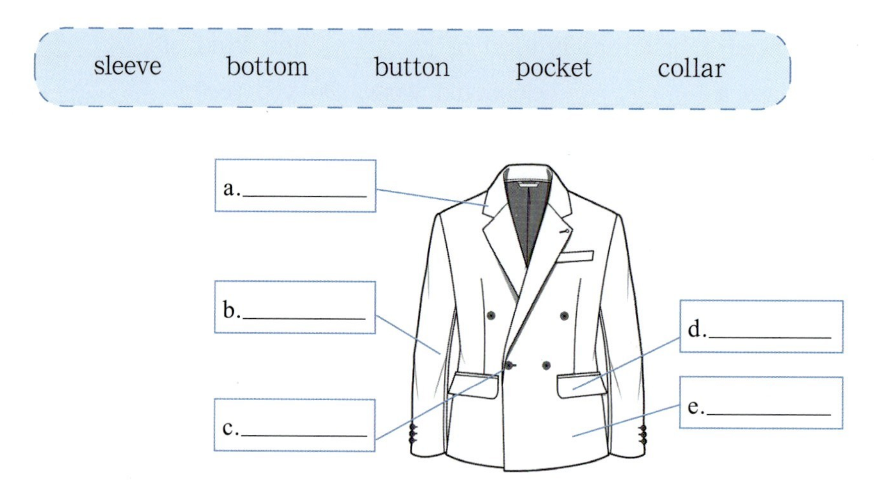

2. Read the passage with the questions.

What collar do you like best? Why?

Garment Components

Collar is a completing upper part of garment. It is seamed on clothes neckline and attached to the neck. It is one of the most important garment components. There are many kinds of collars, such as, square collar, round collar, shirt collar.

square collar

round collar

shirt collar

Sleeve is the part of a garment that covers the arm from the shoulder. The set-in sleeve is the most widely and normally used in design.

Pockets can be either functional or decorative. There are three basic kinds, namely patch pocket, seam pocket and set-in pocket.

Bottom is the lowest part of a garment, skirts, dresses, trousers, etc. It covers five centimeters away from the lowest side of the garment.

Waistband is the top of the trousers and the skirts. It is the part of wearing a belt.

3. Decide true (T) or false (F).

(　　) ①Collar is a upper part of the garment.

(　　) ②Waistband is the bottom of the trousers.

(　　) ③Sleeve covers the shoulder from the arm.

(　　) ④There are only three kinds of pockets.

4. Choose the best answers.

①________ is one of the most important garment components.

A. Sleeve　　B. Pocket　　C. Collar

②Which part is used for wearing a belt? ________.

A. Collar　　B. Sleeve　　C. Waistband

③Bottom covers ________ away from the lowest side of the garment.

A. 4 cm　　B. 5 cm　　C. 15 cm

Tips

set-in sleeve 装袖　　patch pocket 贴袋

seam pocket 接缝口袋　　set-in pocket 嵌入袋

5. Fill in the blanks and try to add the new things you know.

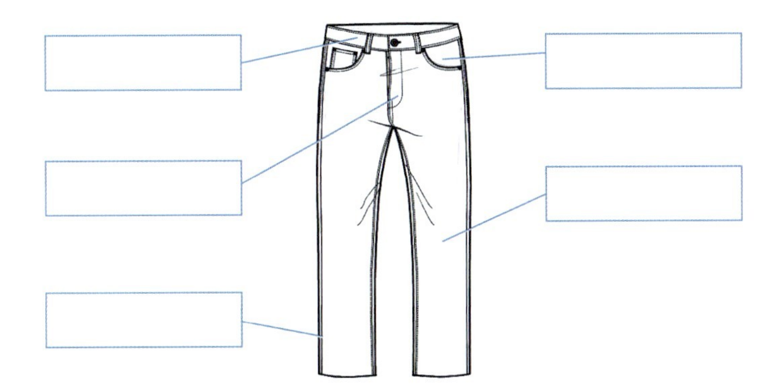

6. Take the measurements with your partner and complete the form below.

Boys

Shoulder: ________　　Waist: ________

Length: ________　　Buttocks: ________

Sleeve: ________　　Leg: ________

Chest: ________

Size: ________　　Size: ________

(The coat you need.)　　(The trousers you need.)

Further Study

1. Look at the pictures below and choose the right number.

Which dynasty do the following garments belong to?

a. Yuan	b. Song	c. Tang	d. Sui
e. Ming	f. Qing	g. Modern	h. Han

☐

☐

☐

☐

☐

☐

☐

☐

☐

☐

☐

☐

2. Choose the best answers.

①Which of the following pictures is the Chinese tunic suit?

A.  B.  C. 

②How many pockets are there in a tunic suit?________.

A. 2 B. 4 C. None

③How many buttons are there on each cuff?________.

A. 2 B. 3 C. 4

④________ is another name of Qipao.

A. Dress B. Cheongsam C. Long dress

⑤Qipao is one of the most typical traditional costumes for Chinese ________.

A. men B. women C. child

⑥Qipao reflects the ________ charm of oriental (东方的) women.

A. elegant B. beautiful C. ugly

3. Write the correct words for the objects in the pictures.

dragon robe embroidery silk

____

____

____

4. Look at the following pictures and say something about the traditional Chinese garments.

Introduction may be as follows:

- Different colors have different meanings in China.
- There are many kinds of garments in China.
- The embroidery is a brilliant pearl in Chinese art.

Sentence patterns may be as follows:

- My favorite color is... in China. It stands for...
- In China, there are many kinds of garments, such as...
- The embroidery is... I like...
- I think the Chinese garments...
- What do you think of Chinese garments?

My Progress Check

1. Words I have learned in this unit are:

□ category	□ standard	□ grade	□ inspector
□ composition	□ discount	□ amount	□ measure
□ waist	□ prefer	□ casual	□ dressy
□ style	□ lighter	□ darker	□ garment
□ component	□ button	□ pocket	□ collar
□ sleeve	□ bottom	□ functional	□ decorative

All together I know ________ words.

More words I know in this unit are:

____

2. Phrases and expressions I have learned in this unit are:

□ safety sort	□ item number	□ original price	□ country of origin
□ look for	□ be interested in	□ would like to	□ belt outlet
□ prefer... to...	□ seam pocket	□ set-in pocket	□ patch pocket
□ be seamed on	□ be attached to	□ many kinds of	□ the top of

Great! Now I know ________ useful phrases and expressions.

More useful phrases and expressions I know in this unit are:

____

3. I can:

□ name some kinds of clothes.

□ understand the sale list.

□ talk with salespeople in the store.

□ identify different parts of a garment.

4. I even can:

□ learn more about the different parts of garment.

□ talk about the traditional Chinese garment with my own words.

UNIT 2

At the Laundry

【Goals】

- Identify different fabrics
- Identify washing labels
- Talk to the clerk(s) at the laundry
- Respond to the customer(s)
- Identify advantages and disadvantages of different fabrics
- Work together to design a garment with Japanese elements

Warming Up

1. Look and match.

a. suit　　b. T-shirt　　c. woolen sweater

d. jacket　　e. jeans　　f. down jacket

g. leather　　h. woolen coat　　i. mink coat

□

□

□

□

□

□

□

□

□

2. Look and write.

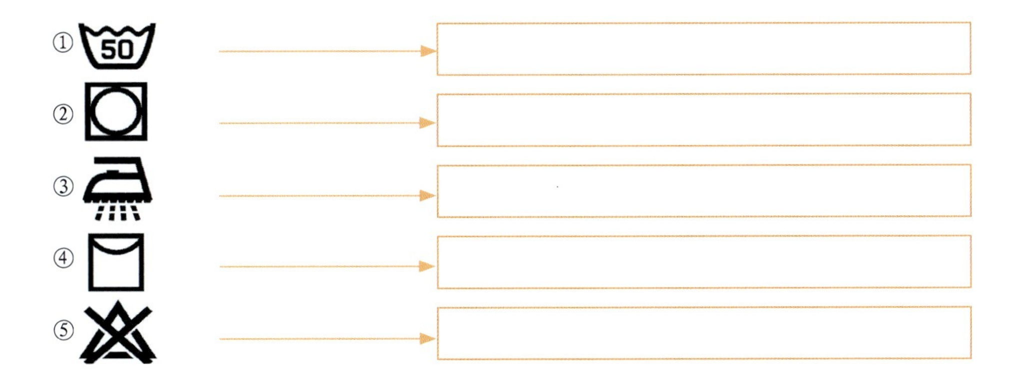

①悬挂晾干　Line Dry
②低温洗涤　Low Wash
③不可折叠　Don' t Bleach
④蒸汽熨烫　Steam Pressing
⑤可转笼干燥　Tumble Dry Menium

3. Draw washing labels as many as you can.

Listening and Speaking

Dialogue One

1. Learn the words.

stain	satisfy	dollar	in total
responsible	damage	get back	have sth. done

2. Listen to the dialogue and repeat.

Clerk: What can I do for you, madam?

Lily: Yeah. I' d like to have this woolen coat dry-cleaned because it has a stain and can' t be washed in water.

Clerk: OK. No problem. We will make you satisfied.

Lily: How much does it cost?

Clerk: 10 dollars in total.

Lily: OK. Are you responsible for the damage?

Clerk: Yes. We are responsible for any damage.

Lily: When can I get my coat back?

Clerk: Tomorrow morning.

3. Listen again and tick the answers.

What did the customer say at the laundry?

☐ What can I do for you, madam?

☐ Are you responsible for the damage?

☐ When can I get my coat back?

☐ I' d like to have this woolen coat dry cleaned.

☐ We will make you satisfied.

☐ We are responsible for any damage.

4. Pair work.

Follow the above example and make a dialogue.

Tips

①I' d like to have this woolen coat dry-cleaned.
我想把这件毛呢大衣拿去干洗。
have sth. done　让……被做，请别人做某事

②We are responsible for any damage.
be responsible for　对……负责
如：You should be responsible for yourself.　你应该对自己负责。

Dialogue Two

1. Learn the words.

pick up	invoice	membership	percent
discount	provide	service	satisfaction

2. Listen and decide whether the sentences are true (T) or false (F).

(　　) ①I come to pick up my laundry with the invoice.

(　　) ②I can get a 30 percent discount next time.

(　　) ③I get a membership card.

(　　) ④I am not satisfied with this service.

3. Listen again and fill in the blanks.

Lily: Hi, I come to pick up my laundry. This is the invoice.

Clerk: OK, ________. I'll get them for you. Here you are. ________ to have a look? Is it to your ________?

Lily: (Look at it) OK, it's very good.

Clerk: This is a ________. You can get a ________ percent discount next time.

Lily: Thanks a lot.

Clerk: Hope to see you again. We will provide ________.

Tips

①pick up my laundry
取回我的衣物

②You can get a 20 percent discount next time.
下一次你可以打八折。

4. Complete the dialogue.

Boss: Good morning, ________________________ (我能为你做些什么), madam?

Tina: Yeah. ________________________ (我有一件非常漂亮的貂皮大衣), but it is a pity that ________________________(上面有一大片油迹) in the sleeves.

Boss: OK. Let me look at it. Don't be worry about it________________________. (这不是一个大问题). We will make you satisfied.

Tina: Oh, thank you! ________________________(洗这件大衣需要多少钱)?

Boss: ________________________ (总共80元人民币).

Tian: OK. ________________________ (什么时候可以拿)?

Boss: Tomorrow afternoon.

Tina: OK, thanks a lot.

Boss: With pleasure.

5. Role play and talk with your partners with the following sentence patterns.

What can I do for you?

I come to pick up my laundry.

Please check your clothes.

You can get a 10 percent discount next time.

How much does it cost?

Reading and Writing

1. Look and discuss.

Can you tell what fabrics are used to make these clothes?

A

B

C

____________ ____________ ____________

2. Read the passage with these questions.

①How many kinds of fabrics do you know?

②What kind of fabrics do you like best? Why?

Various Fabrics

Fabric is the material used to make garments. Fabrics can be classified into cotton, wool and leather and so on. Different fabrics have their own characteristics.

Cotton is often used as a warm-weather fabric because it is light and washable. Cotton is often used in the fashion dress, leisure wear and shirts.

Wool is used for autumn and winter garments because it is heavy, wearable, soft, elegant and warm. It is often used to make high-quality garments. But wool is rather hard to wash or clean.

Leather is usually applied to women's garments and winter garments. It's heavy, warm and graceful. But leather is expensive and hard to store. It requires a lot of care.

3. Decide True (T) or False (F).

( ) ①Fabrics include cotton, wool and leather.

( ) ②Cotton is easy to wash and clean.

( ) ③Leather requires more care than other fabrics.

( ) ④Wool is light and washable so it is often used as a warm-weather fabric.

4. Choose the best answers.

①If you want to make leisure wear and T-shirts, you'd better choose ________.

A. leather B. wool C. cotton

②Which fabric is easy to store and clean? ________.

A. Wool B. Cotton C. Leather

③The characteristics of wool are soft, warm and ________.

A. light B. washable C. wearable

5. Complete the table and try to add the new things you know.

Fabrics	Used to make	Advantages	Disadvantages

Further Study

1. Look at the pictures below and choose the right number.

According to the garments in each picture. Choose the correct country.

a. Thailand	b. Greece	c. India	d. Korea
e. China	f. England	g. Japan	h. Mexico

□

□

□

□

□

□

□

□

2. Look and Write.

Do you know the characteristics of a kimono?

Fill in these blanks below with following words.

collar	clogs	belt	belt junction
Tsunokakushi	cuff(2)		

3. Write the correct words for the objects in the pictures.

sushi	kimono	geta	saki
cherry blossom	Fujisan	sashimi	chopsticks (2)
tatami	treenware	Japanese lantern	

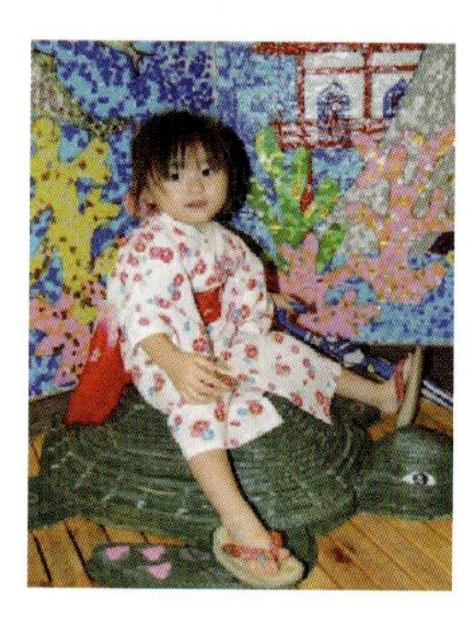

4. Design one of the garments with the elements of Japanese culture.

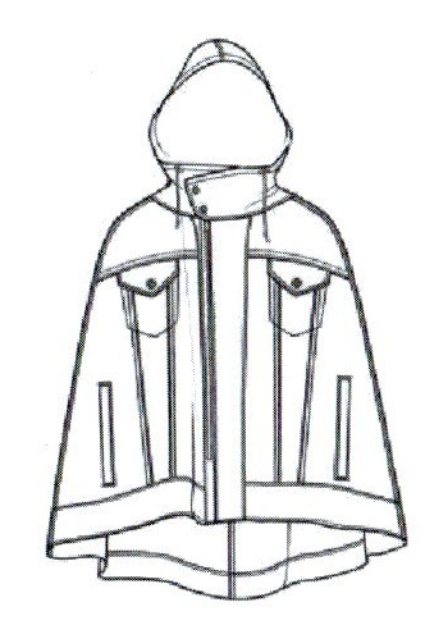

5. Describe your work with a necessary introduction.

Introduction may be as follows:

- Japanese elements in my mind ...
- Types of garments are ...
- Design (fabrics, anime, kimono ...)
- ...

Sentence patterns may be as follows:

- My favorite Japanese element is ...
- This is a ...
- My design focuses on ...
- I use ... for the fabrics because its advantage is ...

My Progress Check

1. Words I have learned in this unit are:

☐ suits
☐ woolen
☐ percent
☐ jeans
☐ down jacket
☐ leather
☐ mink
☐ stain
☐ satisfy
☐ dollar
☐ responsible
☐ damage
☐ invoice
☐ membership
☐ percent
☐ discount
☐ service
☐ satisfaction
☐ various
☐ fabric
☐ characteristics

All together I know ________ words.

More words I know in this unit are:

____

2. Phrases and expressions I have learned in this unit are:

☐ get back
☐ have...done
☐ be responsible for
☐ to one's satisfaction
☐ be classified into
☐ be applied to

Great! Now I know ________ useful phrases and expressions.

More useful phrases and expressions I know in this unit are:

____

3. I can:

☐ identify different fabrics.
☐ identify washing labels.
☐ talk with the clerk(s) at the laundry.
☐ respond to the customer(s).

4. I even can:

☐ identify advantages and disadvantages of different fabrics.
☐ work together to design a garment with Japanese elements.

UNIT 3

In the Market

【Goals】

- Identify different colors and styles
- Identify different colors' healing power
- Talk to the costume designer in the market
- Talk to the customer about alterations
- Learn to provide suggestions on wearing clothes
- Work together to design a garment with American elements

Warming Up

1. Look and match.

a. green　b. red　c. blue　d. purple
e. yellow　f. orange　g. pink　h. black
i. white　j. gray　k. brown　l. gold

Warm Colors

Cool Colors

Neutral Colors

2. Read and translate.

3. Match with colors and their healing power.

red	peaceful
yellow	safe
blue	thoughtful
orange	joyful
green	accommodating
white	gentle
brown	energetic
pink	romantic
purple	bright
black	stable

Listening and Speaking

Dialogue One

1. Learn the words.

attend　　bright　　suit　　tight　　take one's measurement

2. Listen to the dialogue and repeat.

Mary: I am going to attend an important party next month. Please design a dress for me.

Tom: Look at the magazine. Do you like this one?

Mary: It's too bright. The style and the color don't suit me.

Tom: What style do you want?

Mary: I would like a V-neck dress, and it should be tight.

Tom: Oh, I see. How about the color?

Mary: Warm color.

Tom: OK. Do you like this one in the magazine?

Mary: That's what I want.

Tom: Let me take your measurement now.

3. Listen again and tick the answers.

What did the costume designer say in the market?

☐ Do you like this one?

☐ Please design a dress for me.

☐ What style do you want?

☐ How about the color?

☐ The style and color don't suit me.

☐ Do you like this one in the magazine?

4. Pair work.

Follow the above example and make dialogue.

Tips

①The style and color don't suit me. 这个款式和颜色都不适合我。

sth. suit sb. 某物适合某人

Long hair suits her. 留长发适合她。

②Let me take your measurement now. 现在让我帮你量一下尺寸。

take one's measurement 为某人量尺寸

Dialogue Two

1. Learn the words.

try on	alterations	sleeve	take out
put on weight	loose	add	pocket

2. Listen and decide whether the sentences are true (T) or false (F).

(　　)①I can make alternations when they are needed.

(　　)②The sleeves seem a bit too long.

(　　)③Let me take a measurement and make it tight.

(　　)④I' d like to add a pocket here.

3. Listen again and fill in the blanks.

Tom: Please try it on. I can make alterations when they are needed.

Mary: Thank you. (A short while) The ________ seem a bit ________.

Tom: OK. I will ________ the sleeves for you.

Mary: I put on weight these days. So the dress is ________.

Tom: Oh, I see. Let me take a measurement and make it ________.

Mary: Thanks. I' d like to ________here. Could you do it?

Tom: Of course. Anything else?

Mary: No.

Tom: I will notice you when it is finished.

Tips

①I can make alterations when they are needed. 需要的话我可以帮你改一改。

②take out the sleeves 加长袖子 take in 缩短

③V-neck V领 round neck 圆领

4. Complete the dialogue.

Lisa: ______________________ (下个月我要为自己举行一场生日聚会).

Would you please design a dress for me?

Jane: Of course. ______________________ (你喜欢什么款式)?

Lisa: ______________________ (我想要一条圆领的，宽松版的连衣裙).

Jane: OK. What color do you want?

Lisa: ________ (暖色调的). Do you think ____________ (红色适合我吗)?

Jane: Yes. ________________ (现在我帮你量一下尺寸).

5. Role play, talk with your partners with the following sentence patterns.

The dress seems a little bit...

I can... the dress for you.

I lose weight these days.

Do you want to take in the...

Reading and Writing

1. Look and discuss.

Who looks the slimmest?

A

B

C

D

2. Read the passage with these questions.

①Do you know how to look slim?

②What suggestions will you provide to your partner on wearing clothes?

How to Look Slim

STEP 1 wear clothes that fit

I believe that long, oversized tops and slim pants were overdone. Today the first step in a slim look is to wear clothes that are the right size, not too loose and not too tight. Buy what fits and be objective about what length flatters your body type.

STEP 2 use optical iliusions when dressing

Use optical illusion tricks to look slimmer. A-line shapes are great as well as contrast sleeves and color blocking. The black sleeves against the white body are slimming because they frame the white area making it look smaller.

STEP 3 use v-necks, long pants, monochromatic colors, vertical lines

A neckline where you can use jewelry will distract the eye from your mid-section. Longer pants will accentuate your leg length and make you look taller and slimmer.

3. Decide true (T) or false (F).

(　　) ①I believe that long, oversized tops and longer pants were overdone.

(　　) ②The first step in a slim look is to wear clothes that are not too loose and not too tight.

(　　) ③Optical illusion tricks can make you look slimmer.

(　　) ④Wearing jewelry will distract the eye from your mid-section.

4. Choose the best answers.

①What kind of clothes is overdone? __________.

A. Right size　　B. Oversized tops　　C. Longer pants

②_________are slimming because they frame the white area making it look smaller.

A. The white sleeves against the black body

B. The white sleeves against the white body

C. The black sleeves against the white body

③A-line shapes are great as well as contrast __________ and color blocking.

A. sleeves　　B. pocket　　C. waist

④___________can make you look taller and slimmer.

A. Short pants　　B. Slim pants　　C. Longer pants

5. Suppose you are a costume designer, what suggestions will you provide to them?

People	Suggestions
a man with a beer belly	
an obese woman	
a skinny boy	
a petite girl	

6. Answer the questions.

①What's the writer's opinion on wearing oversized tops and slim pants?

②How can we use optical illusion tricks to look slimmer? Please give some examples.

③What's the benefit of wearing a neckline?

Further Study

1. Look and discuss.

Which country do the following pictures belong to?

a. China b. Thailand c. India d. England e. America

2. Look and choose.

a. army uniform	b. swimwear	c. school uniform
d. evening dresses	e. jeans wear	f. POLO
g. sports wear	h. Tailcoat	i. west cowboy

☐

☐

☐

☐

☐

☐

☐

☐

☐

3. Write the correct words in the pictures.

Chocolate	The Eiffel Tower	White House
Statue Of Liberty	NBA	west cowboy
eagle	jeans	Kimono
Great Wall	Harvard University	beefsteak

____

____

____

____

____

____

____

____

____

4. Design one of the garments with the elements of American culture.

5. Describe your work with a necessary introduction.

Introduction may be as follows:

- American elements in my mind...
- Type of garments...
- Colors and tones...
- Design (colors, tones, style...)
- ...

Sentence patterns may be as follows:

- My favorite American element is...
- This is a...
- My design focuses on...
- I use... because its advantage is...

My Progress Check

1. Words I have learned in this unit are:

☐ peaceful	☐ attend	☐ tight	☐ oversized
☐ thoughtful	☐ loud	☐ design	☐ flatter
☐ accommodating	☐ suit	☐ alteration	☐ neckline
☐ energetic	☐ distract	☐ loose	☐ pants
☐ romantic	☐ jewelry	☐ pocket	☐ measurement
☐ bright	☐ accentuate	☐ slim	☐ add

All together I know ________ words.

More words I know in this unit are:

____

2. Phrases and expressions I have learned in this unit are:

☐ warm color	☐ color blocking	☐ take out	☐ vertical lines
☐ cool color	☐ A-line shapes	☐ put on weight	☐ long pants
☐ bottle green	☐ V-neck dress	☐ make alternation	☐ design ... for
☐ charcoal gray	☐ try on	☐ optical illusion	☐ suspender pant

Great! Now I know ________ useful phrases and expressions.

More useful phrases and expressions I know in this unit are:

____

3. I can:

☐ identify different colors and tones.

☐ identify different colors' healing power.

☐ talk to the costume designer in the market.

☐ talk to the customer about alteration.

4. I even can:

☐ learn to provide suggestions on wearing clothes.

☐ work together to design a garment with American elements.

UNIT 4

In the Negotiation

【Goals】

- Identity different ways of shipment
- Learn words and patterns on taking and making orders
- Communicate with the customers as a garment merchandiser about payment terms
- Master the forms of purchase orders and try to make an order with your partner(s)
- Know about some French garments in a fashion show

Warming Up

1. Look and match.

a. free shipping
b. ground transportation
c. handle with care
d. air transport
e. shipping cart
f. water transport
g. shipping in 24 hours
h. green transport
i. keep dry
j. freight check-in
k. this side up
l. freight claim

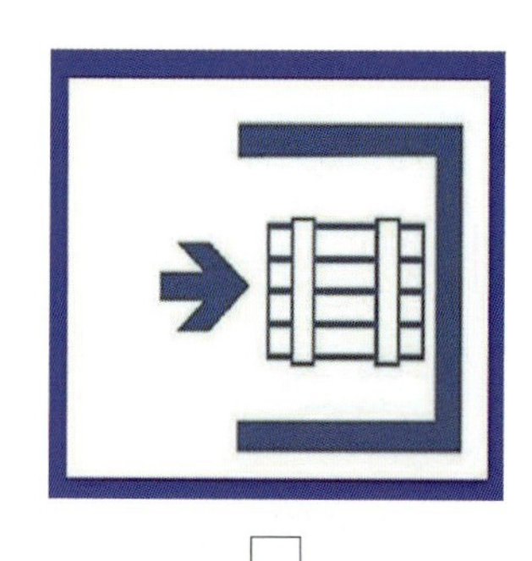

2. Look and translate.

Do you know about some lists of services/products?

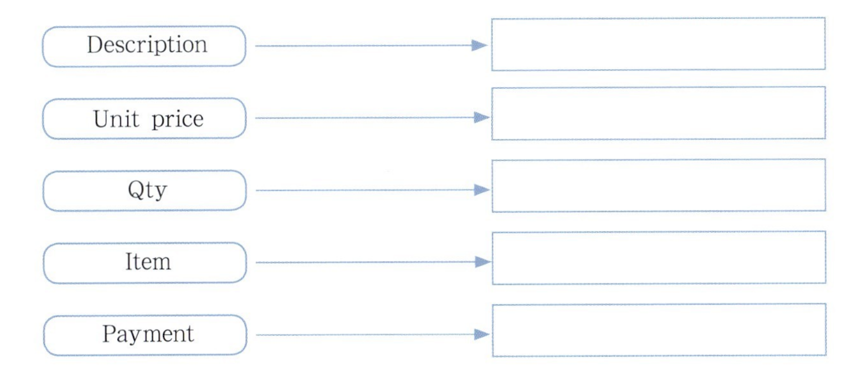

3. Follow the example and draw a price tag with information as much as you can.

Listening and Speaking

Dialogue One

1. Learn the words.

quality dear come down recommend handmade bargain

2. Listen to the dialogue and repeat.

Customer: Please show me your new product.

Salesclerk: Yes, here you are.

Customer: It's in good quality. What about the price?

Salesclerk: Sixty *yuan*.

Customer: Oh, it's too dear. Can you come down a bit?

Salesclerk: Sorry, but this is the best price. You can't get it any cheaper than here.

Customer: Are you sure?

Salesclerk: Of course.

Customer: Could you recommend products good for sale?

Salesclerk: Sure. How about this style? They are all handmade.

Customer: Oh, it's good. How much is it?

Salesclerk: It's forty *yuan*.

Customer: OK, I'll take 100 pieces.

3. Listen again and tick the answers.

What did the customer say about the bargain?

☐ What about the price?

☐ It's too dear. Can you come down a bit?

☐ Sorry, but this is the best price.

☐ How much is it?

☐ It's forty *yuan*.

☐ OK, I'll take 100 pieces.

4. Pair work.

Follow the above example and make a dialogue.

> **Tips**
>
> ①It's in good quality. 它的质量很好。 in good quality 质量好的
>
> The clothes are in good quality.
>
> ②Could you recommend products good for sale? 你可以推荐好卖的产品吗?

Dialogue Two

1. Learn the words.

order   sample   profit   deliver to sb.   the same as before

2. Listen and decide whether the sentences are true (T) or false (F).

( ) ①Jim wants to order the same style of last time.

( ) ②Carl will go to have a look at some new samples.

( ) ③This time Carl orders at the same price as before.

( ) ④At last they make an order of 1,000 pieces.

3. Listen again and fill in the blanks.

Jim: Hello! This is Jim. Can I help you?

Carl: Yes. ______ (我想订上次那款货).

Jim: We have new samples. Come to have a look.

Carl: I'm not free.

Jim: OK. You want the style of last time?

Carl: Yes, how much? ______ (和以前一样价吗)?

______ (这次我多要些，便宜点). OK?

Jim: No, ________ (上次的价格已经是最优惠的了). Impossible! Believe me. We have no profit every time. We are good friends. OK. ________ (要多少件)?

Carl: One thousand. Call me when ready.

Jim: You take or we deliver to you?

Carl: I will call you. OK, goodbye!

Jim: Goodbye!

Tips

①the style of last time 上次的款式
②How many pieces? 多少件?
③Call me when ready. 准备好了就给我打电话。

4. Complete the dialogue.

cooperation	trial order	quoting process	payment terms
currency	rate	TT	L/C

Mike: Hi, Simon. ________ (订单考虑得怎样了)?

Simon: Since this is the first cooperation, I would place 10,000 pcs. as trial order. And ________ (迟点我会把具体的样板、款式以及订单发给你). Now I would like to know as per our quoting process and payment terms, to see if it's OK for you.

Mike: OK. What currency to use for quoting?

Simon: Hong Kong dollar, please.

Mike: Great! ________ (报价使用哪天的汇率)?

Simon: As per the date you quote.

Mike: OK. What's more, ________ (能否考虑将汇款方式改为电汇呢)? If to use L/C, charges are higher.

Simon: But it would be ________ (使用信用证付款方式比较保险) for the first cooperation.

Mike: No problem. After you can send us relative materials, we will arrange a merchandiser to follow up and assist you. ________ (期待能和你建立长期的合作关系).

Simon: Thank you, and I hope so.

5. Role play and talk with your partners with the following patterns.

I would like to order...

This time I order more, cheaper, OK?

Sorry, last price already, and we have no profit.

I would place 10,000 pcs. as trial order.

What currency to use for quoting?

Reading and Writing

1. Tick and discuss.

Can you tick which item belongs to the part of an order?

PO Number ________

Shipping Date ________

Sample ________

Terms of Payment ________

Billing Address ________

2. Read the passage with these questions.

①How many kinds of size do you know?

②Which do you care most during the purchase of order? Why?

JJ's Store Purchase Order Form

Quantity (pcs.)	Description of Goods	Color	Unit Price	Price	Size		
					M	L	XL
800	silk gown (Item No. SG10)	black	$10.00	$8,000.00	200	500	100
1,200	ladies' tights (Item No. LT20)	Blue	$6.00	$7,200.00	500	500	200
1,000	bra (Item No. B30)	white	$5.00	$5,000.00	400	400	200
1,000	pajamas (Item No. P40)	pink	$8.00	$8,000.00	600	300	100

silk gown 丝绸长袍　ladies' tights 女式紧身衣　bra 胸罩　pajamas 睡衣

3. Choose the best answers.

①How many kinds of clothes does the customer order? ________.

A. 2　B. 3　C. 4

②What's the unit price of the silk gown? ________.

A. $10　B. $5　C. $8

③Which kind of clothes is the cheapest? ________.

A. Silk gown　B. Bra　C. Pajamas

④What's the total quantity of the order? ________.

A. 4,000　　B. 28,200　　C. It's not mentioned

4. Complete the purchase order with information you like.

JJ's Store is going to purchase some underwears from Company FOX.

PURCHASE ORDER

JJ's Store

____

____

____

PO #________

Date: ________

To (address)

____

____

____

Tel: ________________

Fax/Email: ________________

Ship to (address)

____

____

____

Tel: ________________

Fax/Email: ________________

Comment　150 characters left

Quantity	Description	Unit Price	Amount

	Subtotal:
	Sales Tax: 9%
	Shipping & Handing
	Total Due:

Make all checks payable to company name.

Total due in 15 days, overdue accounts subject to a service charge of 5% per month.

THANKS FOR YOUR BUSINESS!

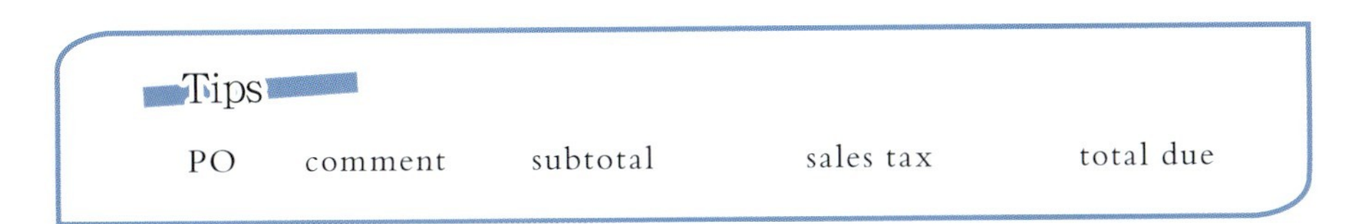
Tips

PO   comment   subtotal   sales tax   total due

5. Translate the order and then make up a similar one according to the provided information.

①Please send to the following items to be shipped by way express, and bill us.

②The order is contingent on receiving the terms of 2% within 30 days: 800 pcs. silk gowns (black); 1, 200 pcs. ladies' tights (blue); 1, 000 pcs. bras (white); 1, 000 pcs. pajamas (pink).

③Please send to the __________ to be shipped by way express and bill us. The order is contingent on __________ the terms of 2% within 30 days: 500 pcs. ______ (blue); ____________ T-shirts (________); ______________ (________).

(500 pcs.)   (800 pcs.)   (600 pcs.)

Further Study

1. Look and choose.

Match the following pictures with the different periods.

a. The Baroque　　b. The Rococo　　c. The Emperor　　d. Modern Times

□

□

□

□

□

□

□

□

□

2. Look and Write.

Do you know the characters of the garments?

Fill in blanks with the following words.

puff sleeve   lace shirt   hat   shorts to knee   loose-fitting

3. Write the correct words for the objects in the pictures.

irises	perfume	whale bone skirt
foie gras	the Eiffel Tower	national flag
shorts to knee		

________________ ________________ ________________

________________ ________________ ________________

4. Design one of the garments with the elements of French culture.

5. Describe your work with a necessary introduction.

Introduction may be as follows:

- Paris Fashion Show in my mind...
- Type of garments...
- Design (fabric, color, perfume...)
- ...

Sentence patterns may be as follows:

- My favorite French element is...
- This is a...
- My design focus on...
- I use... for... because...
- ...

My Progress Check

1. Words I have learned in this unit are:

□ description	□ quality	□ item	□ payment
□ dear	□ recommend	□ handmade	□ bargain
□ order	□ sample	□ profit	□ deliver
□ cooperation	□ currency	□ rate	□ TT
□ L/C	□ quantity	□ bra	□ pajamas
□ PO	□ comment	□ subtotal	□ express

All together I know ________ words.

More words I know in this unit are:

____

2. Phrases and expressions I have learned in this unit are:

□ free shipping	□ ground transport	□ handle with care	□ air transport
□ shipping cart	□ water transport	□ unit price	□ payment terms
□ trial order	□ freight claim	□ quoting process	□ billing address
□ total due	□ be contingent on	□ shipping date	□ list of service

Great! Now I know ________ useful phrases and expressions.

More useful phrases and expressions I know in this unit are:

____

3. I can:

□ identify different ways of shipment.

□ identify words and patterns on taking and making orders.

□ communicate with the customers about payment terms.

□ master the forms of purchase order and try to make an order with my partner(s).

4. I even can:

□ know about some French garments in Fashion Show.

□ design a garment with French elements.

UNIT 5

In the Workshop

【Goals】

- Learn about names of basic clothing machines, equipment and basic production flow of garments
- Talk to customers about goods packing
- Introduce the clothing workshops to the customer
- Learn to confirm and send orders of clothes purchase
- Know about basic information about Indian garments
- Work together to design a garment with Indian elements

Warming Up

1. Look and match.

a. thread　　b. iron　　c. tape measure

d. sewing machine　　e. shuttle　　f. chalk

i. thimble　　J. tailor' s scissors　　k. manual awl needle

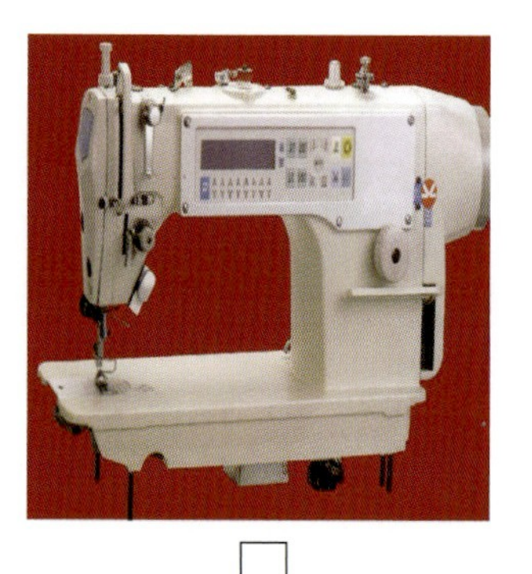

☐

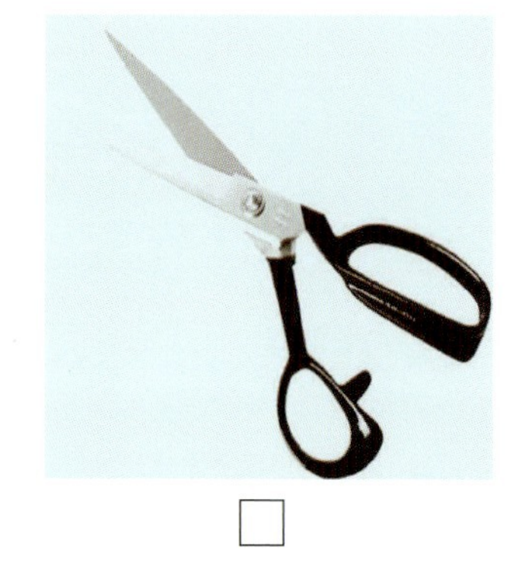

☐

☐

☐

☐

☐

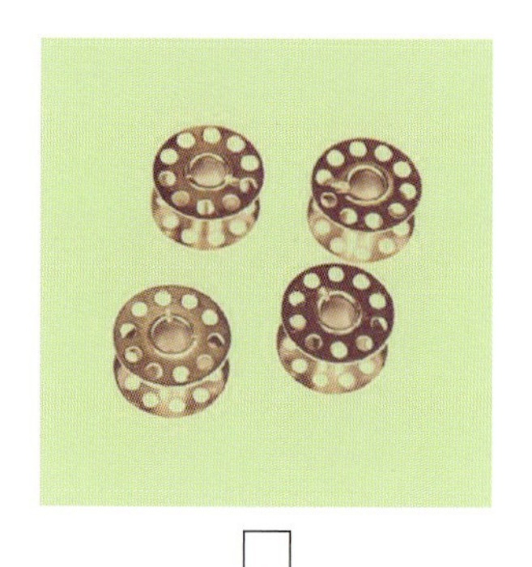

☐

☐

☐

2. Look and translate.

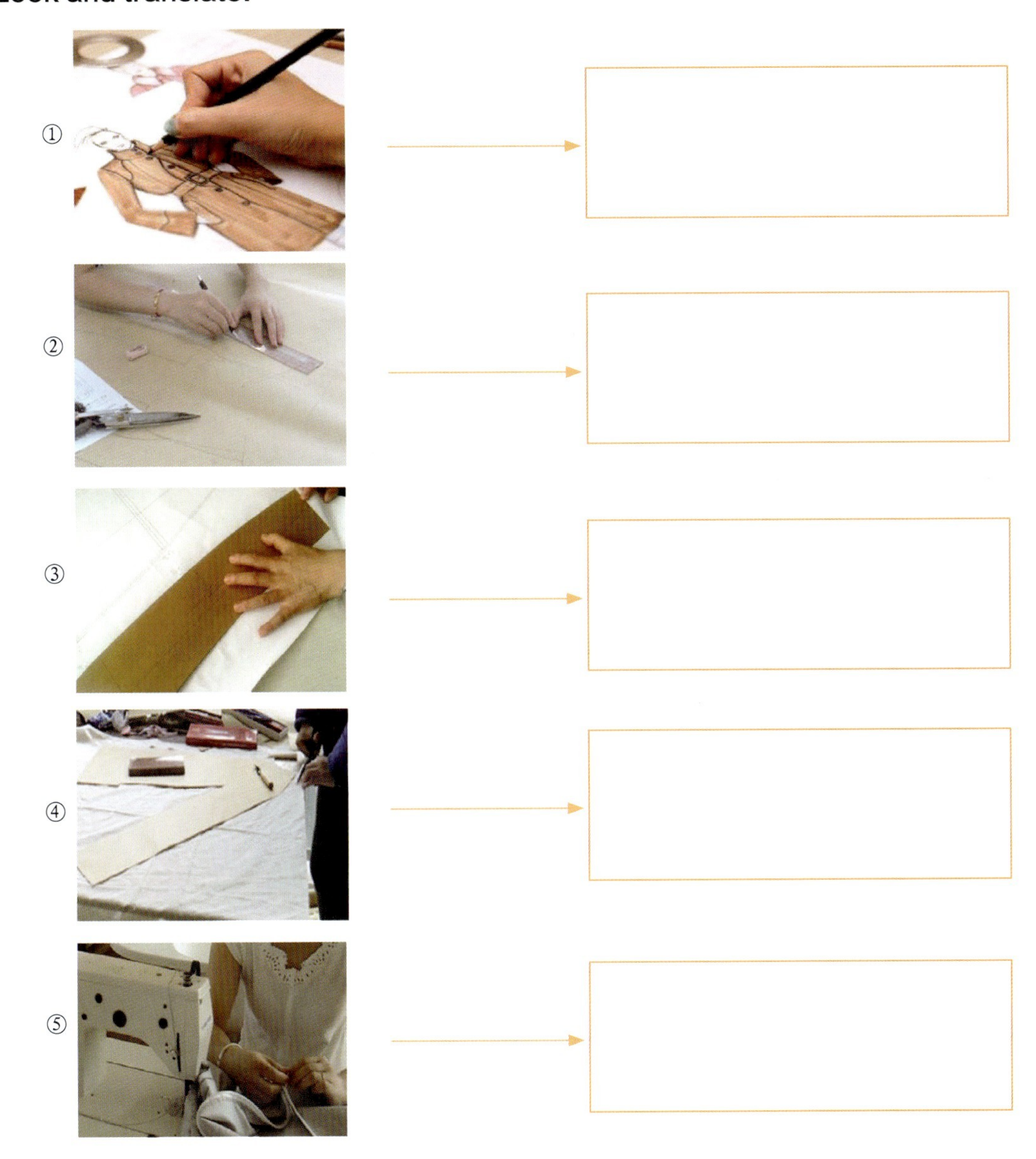

3. Try to arrange the order of production flow according to your common sense.

a. Bulk Goods Production
b. Delivery Goods
c. Quotation
d. Pattern Making
e. Receive the Order

______ → ______ → ______ → ______ → ______

Listening and Speaking

Dialogue One

1. Learn the words.

packing   requirement   inner   outer   carton

2. Listen to the dialogue and repeat.

Manager: How about packing? Do you have any requirements on it?

Customer: We require inner packing and outer packing.

Manager: How about inner packing?

Customer: Plastic paper bag for each piece will be OK.

Manager: How about outer packing?

Customer: Each 50 pieces in a carton.

Manager: OK, according to your requirements.

3. Listen again and tick the answers.

What did the merchandiser say about the packing?

☐ How about packing?

☐ Do you have any requirements on it?

☐ Have inner packing and outer packing.

☐ Plastic paper bag for each piece will be OK.

☐ How about outer packing?

☐ OK, according to your requirements.

4. Pair work.

Follow the above example and make a dialogue.

①Do you have any requirements on it ?　请问你对包装有什么要求?
②Plastic paper bag for each piece will be OK.　每件用塑料袋包装就行。
③Each 50 pieces in a carton.　每箱50件。

Dialogue Two

1. Learn the words.

opportunity	mainly	manufacture	evening dress
bridal gown	employ	workshop department	similar

2. Listen and decide whether the sentences are true (T) or false (F).

(　　) ①The factory doesn' t produce evening dress but bridal gowns.

(　　) ②There are less than one thousand engineers in the factory.

(　　) ③There are mainly four departments in the factory.

(　　) ④Their prices are much higher than those of the similar products.

3. Listen again and fill in the blanks.

Manager: Welcome to our factory, Mr. Peter.

Peter: I' m glad to have the opportunity to visit your factory. I' ve heard a lot about it.

Manager: Thank you. ________________________________.

Peter: I' m told that you mainly manufacture evening dress.

Manager: You' re right. ________ of our products are evening dresses and the rest are bridal gowns.

Peter: How many engineers do you employ?

Manager: There are about ________ engineers.

Peter: How many workshops do you have?

Manager: There are mainly four ________. They are Cutting Design, Cutting, Sewing and Finish in workshops.

Peter: I know that your prices seem to be 5% higher than those of the ________ products.

Manager: Yes, our prices are ________ higher, but the quality is of first class.

Peter: I see. Thank you for taking me around, Mr. Wang.

Manager: It's my pleasure!

Tips

①Your prices seem to be 5% higher than those of the similar products.
你们的价格好像比同类产品高5%。

②The quality is of first class.
质量是一流的。

4. Complete the dialogue.

Rose: Hello, Mr. Jacky. I'm Rose. Glad you could come.

Jacky: ______________________________. (感谢你邀请我来参观工厂。)

Rose: Where shall we start?

Jacky: ______________________________? (能先谈谈车间吗?)

Rose: OK. This is the Cutting Design workshop. It is a ________ (合资企业) with an Italian corporation.

Jacky: This workshop looks clean and bright.

Rose: Yes. It is designed to create satisfactory ________ (工作环境).

Jacky: Do you make evening dresses?

Rose: Yes. It's a fashion nowadays among Europeans and Americans to admire hand-made things.

Jacky: ______________________________? (你认为晚礼服在中国的前景如何?)

Rose: I think they ________ (市场潜力很大). I am confident that their demand will exceed their supply in a few years.

Jacky: This must bring you ____________ (丰厚利润)。

Rose: ____________ (我希望如此). Shall we go to the sales hall to have a look?

Jacky: Let's go.

5. Role play and talk with your partners with the following patterns.

Welcome to...

I'm glad to have the opportunity to visit your factory.

How many engineers / workshops do you have?

Do you make / produce / manufacture...?

Thank you for taking me around.

Reading and Writing

1. Look and discuss.

Can you tell which department they belong to?

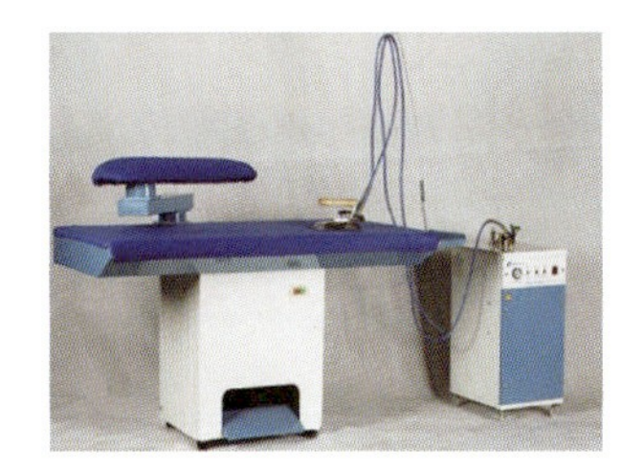

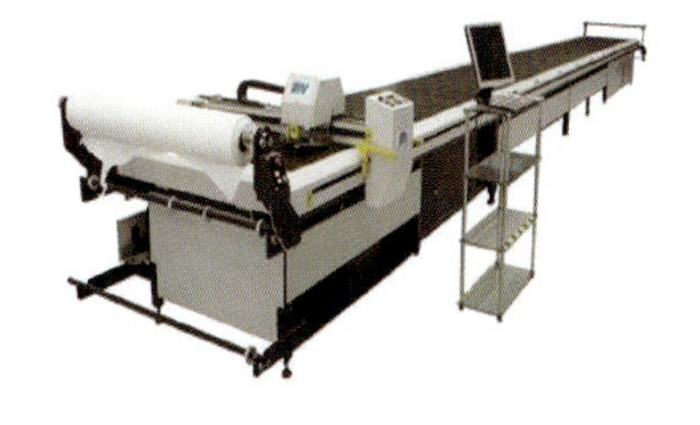

a. ________ b. ________ c. ________

2. Read the passage with these questions.

①How many clothing workshops do you know?

②Which department do you care most for the purchase of order? Why?

There are many departments in the Production Department, such as Merchandising Room, Pattern Room, Fabric and Accessories Warehouse, Cutting Room, Sewing Room, Pressing Room and Packing Room, etc. For the OEM (Original Equipment Manufacturer), the merchandisers will ensure smooth production and timely delivery, such as dealing with orders, purchasing fabric and accessories, contacting with the customers, making the production sheet, etc.

Here in the Pattern Room, pattern designing, grading, sample making are finished. The designers will work closely with the workmanship team. Those lower sections are in the same grade and charged by the Production Department.

This is the Fabric and Accessories Warehouse. The merchandisers and the Technology Team Director will work out how many kinds of fabric and how much fabric does

every order and every style cost. Then they'll fill the information in the purchasing sheet; give it to the fabric buyer. At last the workshop will fetch it for production.

In the Cutting Room, the workers will do the fabric spreading and marker planning as per the production sheet. They will plan it well before they cut the fabric into pieces of garment. It's a tough job because the workers need to walk a lot around the cutting board every day.

3. Decide true (T) or false (F).

( ) ①Merchandisers should ensure smooth production and timely delivery.

( ) ②Only pattern designing is finished in the Pattern Room.

( ) ③The merchandisers buy fabric after filling in the purchasing sheet.

( ) ④The workers do the fabric spreading and marker planning in the Cutting Room.

4. Choose the best answers.

①If you want to make orders, you'd better contact with ________ first.

A. Merchandising Room　　B. Cutting Room　　C. Pattern Room

②Who will be in charge of the merchandising of OEM? ________.

A. Production Department　　B. The merchandiser　　C. OEM

③Fabric purchasing sheet will be counted by ________.

A. Merchandising Room　　B. Fabric and Accessories Warehouse

C. Pattern Room

5. Translate the following sentences.

①This is to confirm my telephone order of yesterday for the following items: 4 Jr. Sewing Machines Model 3A; 7 Homemaker's Ironing Boards; 15 Fold Up Clothes Racks.

____

____

____

②Thank you for your samples of woolen coats received today. Please make a shipment in accordance with our Order No. 2602 enclosed here with.

____

____

____

③Please confirm the delivery schedule with supplier after you place the order.

____

____

____

Further Study

1. Look and choose.

Can you name them quickly?

a. nose ring	b. saris	c. bloomer
d. kurta	e. choli	f. earings
g. dhoti	h. turban	i. hemming full-length skirt

☐

☐

☐

☐

☐

☐

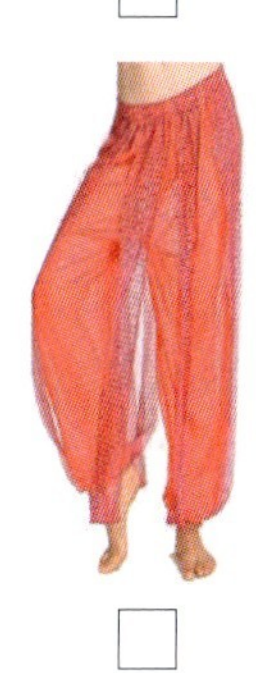
☐

☐

☐

2. Look and Write.

Do you know the characters of sari?

Fill in the blanks with correct words.

saris (2) hand chain choli hollow nail diamond bangles henna

3. Write the correct words for the objects in the pictures.

peacock skirt lotus head chain Taj Mahal
hand pilaf ornament case necklace and earrings

______

______

______

______________________ ______________________ ______________________

4. Choose some hennas and design one garment of Indian culture.

5. Describe your work with a necessary introduction.

Introduction may be as follows:

- Indian elements in my mind...
- Type of clothing machines and equipment...
- Design (color, fabrics, hennas...)
- Saris in use...

Sentence patterns may be as follows:

- My favorite Indian element is...
- This is a...
- My design focuses on...
- I use... for the saris (skirt...) because...

My Progress Check

1. Words I have learned in this unit are:

☐ thread ☐ sewing needle ☐ tailor's scissors ☐ chalk
☐ button ☐ shuttle ☐ thimble ☐ bobbin
☐ packing ☐ carton ☐ iron ☐ inner
☐ outer ☐ employ ☐ requirement ☐ similar
☐ opportunity ☐ mainly ☐ manufacture ☐ quotation

All together I know ________ words.

More words I know in this unit are:

____

2. Phrases and expressions I have learned in this unit are:

☐ bulk goods production ☐ deliver goods ☐ evening dress
☐ bridal gown ☐ pattern making ☐ take around
☐ first class ☐ Workshop Department ☐ contact with
☐ fabric spreading ☐ deal with ☐ make shipment
☐ in accordance with ☐ Production Department ☐ confirm orders
☐ production sheet ☐ manual awl needle ☐ tape measure
☐ sewing machine

Great! Now I know ________ useful phrases and expressions.

More useful phrases and expressions I know in this unit are:

____

3. I can:

☐ identify some basic clothing machines and equipment.
☐ identify basic production flow of garments.
☐ talk about goods packing with customers.
☐ introduce the clothing workshops to the customer.

4. I even can:

☐ learn to confirm and send orders of clothes purchase.
☐ know about basic information about Indian garments.
☐ work together to design a garment with Indian elements.

UNIT 6

In the Trade Fair

【Goals】

- Identify types of booth
- Talk about booking a booth
- Respond to booth booking
- Communicate with customers in the trade fair
- Compare the methods of clothing display
- Understand the national costume about England and Scotland

Warming Up

Look and match.

a. vendor booth b. fashion show c. craft fair
d. showroom e. corner booth f. pop-up store
g. tent circle booth h. vintage market i. fashion truck

☐

☐

☐

☐

☐

☐

☐

☐

☐

Listening and Speaking

Dialogue One

1. Learn the words.

garment	booth	style	designing house
reserve	location	discount	according to

2. Listen and choose.

①What does the man want to do?

A. Book a booth.

B. Buy a suit.

C. Attend the fair.

②What booth style does he want?

A. Corner booth.

B. Island booth.

C. Row booth.

3. Listen and complete.

Grace: Good morning. May I help you?

Jack: Good morning, I would like to ________ the fashion fair in July! And can I reserve (预定) a booth?

Grace: Sure. We have island ________, corner booth, row booth. What type of booth would you like?

Jack: Em, I'd like to book a corner booth. What is the ________ of a booth?

Grace: The price is set according to the size and location (位置). The corner booth is ________.

Jack: I see. But I think the price is too high.

Grace: Em, I will give you 5% ________.

Jack: You are helpful.

Grace: By the way, here are some forms you need to ________.

Jack: OK! When should I ________?

Grace: Before April ________.

Jack: Thank you. You are very nice. See you.

Grace: Our pleasure. Bye.

4. Decide true (T) or false (F).

( ) ①The exhibition will be held in July.

( ) ②At last, Jack booked a corner booth.

Dialogue Two

1. Learn the words.

detail	participate	submit	catalog
island booth	corner booth	row booth	end booth
exhibition	exhibitor	billboard	

2. Listen and repeat.

Carl: Good morning.

Joan: Good morning. I' m Joan, the sales manager from Company Fox.

Carl: How may I help you?

Joan: I heard that an exhibition will be held in April. Could you tell me more details about it?

Carl: Sure. We have island booth, corner booth, row booth, end booth. And this is our catalog.

Joan: Thanks. Em, I' d like to book a corner booth, about normal size, but I think the price is too expensive.

Carl: Is this the first time you participate in our exhibition?

Joan: Yes!

Carl: Em, we have a 5% discount for new exhibitor.

Joan: Oh! I see. That would be better than before.

Carl: And here are some forms you need to fill in.

Joan: Thank you. When should I submit?

Carl: Before January 30th.

Joan: As you know, it is our first time to participate in the exhibition. What should we pay attention to?

Carl: You should attract your visitors by your booth design.

Joan: Sure. Can I arrange some people to hold a billboard, and walk in the exhibition?

Carl: It is also a good idea.

Joan: Em, I see. Thanks for your support.

Carl: Our pleasure. Thank you for your coming. Bye.

Joan: Bye.

3. Listen again and answer the questions.

What type of booth does the sales manager want?	
How many times has the sales manger attended the fair?	
When should she submit the forms?	

4. Read and order.

Write down numbers to make sure the following sentences are in the correct order.

(　　) What can I do for you?

(　　) You' re welcome.

(　　) 30,000 *yuan*.

(　　) How much is it?

(　　) OK, I' ll take it. Thank you.

(　　) I want to reserve a booth.

(　　) What type do you want?

(　　) Island booth.

5. Pair work.

Follow the above example and use the sentence patterns below to make a dialogue.

organizing committee: 组委会

exhibitor: 参展商

May I help you?
What kind of booth would you like...?

I'd like to book...
What's the price of...?
When should I submit...?

Reading and Writing

1. Choose and write.

fashion fair	contrast display	garments
book	display	customers

a. Jack's company wants to attend the ________.

b. Jack goes to ________ booth.

c. Jack designs the ________.

d. Jack uses the ________.

e. There are many ________ to order their products.

f. After the fashion fair, Jack began to pack their ________.

2. Read the passage with these questions.

How many methods of display are mentioned? What are they?

Display Methods

The aim of clothing display is to establish a brand image and promote the sales. The excellent display can attract customers, and can help the company to meet the market profit maximization. We should use artistic techniques to highlight the characteristics of the clothing and selling point. Try our best to maximize customers' purchase desire. So before displaying clothing, we should consider the following five basic points: ①visible; ②easy to touch; ③easy to select; ④easy to buy; ⑤having rich or meaningful personality.

In order to promote product sales, we should pay attention to the methods of display, especially to those listed as below:

(1) Contrast display

This method uses the collocation of color, light and shade to form bright contrast, which can have a strong attraction. It also can give a person deeper impression.

(2) Symmetry display

This method is suitable for a large quantity of goods. It can give a person a sense of security, and a sense of balance.

(3) Classified display

According to a certain level of category, we can pack our goods in one or two classifications. This is a method of a relatively concentrated display.

(4) Theme display

With a specified date or festival as the theme, we can make the display of goods associated with the theme, and this can help us to adapt to the public psychology. It also can create the sales atmosphere.

(5) Season display

According to the climate, seasonal changes, we can make the seasonal goods into the real time display. In this way, we can meet the customers' spending habits.

When we design a display, we can combine the lamination with mount. Lamination means that we should stack the clothing one by one. Generally through the orderly and coordinated clothing folding, we can display the goods on the goods shelf. This way can save the limited space. But its disadvantage is that it is not the method to present the products completely.

Therefore, it must be adjusted to mount shows. Mount, it is generally with a coat hanger to hang clothes on, so as to fully show the products' characteristics. It is easy to form a visual color impact and render the atmosphere, so that consumers can know the goods directly. However, in the limited store, it is not possible to hang up the display that the best combination is mounted with a stacking. In this way, we can reasonably use the space. On the other hand, we also make the whole commodity display, and have a sense of hierarchy. It can promote our products sales.

3. Look and match.

a. Season display　　b. The collocation of lamination and mount

c. Mount display　　d. Lamination display

__________　__________　__________　__________

4. Decide true (T) or false (F).

(　　) ①The five basic points for displaying clothing help to highlight the characteristics of the clothing and selling point.

(　　) ②There are 4 methods of displaying clothing.

(　　) ③Mount display can help the customers know the goods directly.

(　　) ④The best way of display is to combine the mount with lamination.

5. Read and complete the form.

Purpose of display	
Five basic points	①______ ②______ ③______ ④______ ⑤______
Methods of display	①______ ②______ ③______ ④______ ⑤______
Advantages of lamination display	

Further Study

1. Look and choose the right letter.

a. Scottish kilt	b. The Queen's hat	c. British Flag
d. Big Ben	e. Victoria costume	f. London Bridge
g. British Royal Family	h. Tail coat	i. Buckingham Palace

□

□

□

□

□

□

2. Look and write.

Do you know the characters of Victorian costume?

Fill in the blanks with the correct words.

bows　collar　high waist　lotus leaf

multi-layered cake cutting　leg of mutton sleeves

3. Match the pictures to coutries.

Where did each couple have their wedding according to their wedding clothes or wedding traditions?

a. Wales　b. Scotland　c. England

d. Ireland　e. Northern Ireland

□

□

□

4. Design a hat with the elements of British culture.

5. Describe your work with a necessary introduction.

Introduction may be as follows:

- British elements in my mind...
- Type of garments...
- Colors and tones...
- Design (colors, tones, style...)

Sentence patterns may be as follows:

- My favorite British element is...
- This is a...
- My design focuses on...
- I use... because its advantage is...

My Progress Check

1. Words I have learned in this unit are:

☐ booth ☐ exhibitor ☐ garment ☐ reserve
☐ location ☐ discount ☐ catalog ☐ submit
☐ billboard ☐ display ☐ promote ☐ highlight
☐ collocation ☐ coordinated ☐ stack ☐ mount
☐ concentrated ☐ display

All together I know ________ words.

More words I know in this unit are:

____.

2. Phrases and expressions I have learned in this unit are:

☐ according to ☐ visual impact ☐ contrast display
☐ designing house ☐ brand image ☐ theme display
☐ symmetry display ☐ classified display ☐ associated with
☐ adapt to ☐ island booth ☐ corner booth
☐ row booth ☐ fashion fair

Great! Now I know ________ useful phrases and expressions.

More useful phrases and expressions I know in this unit are:

____

3. I can:

☐ identify booth types.
☐ talk about booking a booth.
☐ compare the methods of clothing display.

4. I even can:

☐ communicate with customers in the trade fair.
☐ understand the national costume about England and Scotland.

UNIT 7

In the Electronic Commerce (E-commerce)

【Goals】

- Identify tools using in the E-commerce
- Identify details in product catalog
- Talk about products online
- Get general ideas of online customer services
- Make suggestions to deal with problems about online shopping
- Talk about Egyptian elements with your partner

Warming Up

1. Look and match.

a. AI	b. cash	c. CAD
d. computer	e. smart phone	f. Coreldraw
g. Alipay	h. credit card	i. Photoshop

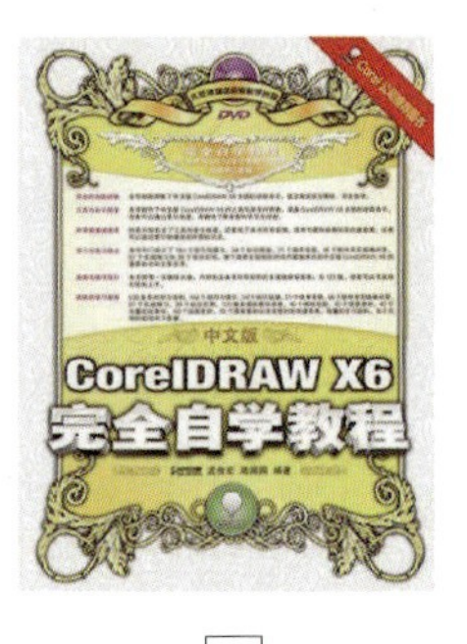

2. Tick as much information as you can find on the catalog.

Wholesale

+1 (213) 488-0226
Global Offices:

Catalog

Search Products
Style # / Name Find

Express Order 2.0
E-Mail Address / Username
Password
Lost Password? Login
Need an Account?

Wholesale Catalog
- New & Limited
- Women
- Men
- Kids
- Babies
- Accessories
- Bags
- Organics
- NEON
- Sublimation

Resources
Global Offices
Distributors

Lavender

2102
Fine Jersey
Short Sleeve Women's T

This classic fitted T is our most popular style for Women. Ultra soft and smooth 100% Fine Jersey Cotton provides a great surface for embellishment and makes the perfect everyday T.

Size Chart Specifications

- 100% Fine Jersey cotton construction (Heather Grey contains 10% Polyester)
- Durable rib neckband
- Form-fitting
- Fabric weight 4.3 oz/yd² (146 g/m²)
- Superior screen printing results

Made of 100% fine ring-spun combed cotton, this lightweight fine jersey is exceptionally smooth and tight-knit, making it a perfect surface for screen printing.

See other versions

☐ gender	☐ color	☐ price
☐ specifications	☐ fabric	☐ style
☐ size chart	☐ products catalog	☐ global offices

3. Look and answer.

①What product is this? ______________________________

②How many colors are there? ______________________________

③What's the number of the T-shirt? ______________________________

④Is it for the young? ______________________________

⑤What fabric is it? ______________________________

2201 ShortSleeve T-Shirt
Youth
Fine Jersey

Listening and Speaking

Dialogue One

1. Learn the words.

service	stock	expensive	discount
profit	turnover	delivery	arrival

2. Listen to the dialogue and repeat.

Customer: Hello! Is anybody here?

Customer Service Agent: Hello! Welcome to our shop. Glad to be of service, what can I do for you?

Customer: I like this style of dress.

Customer Service Agent: My dear! There are enough stock of this commodity, so you can book ahead and pay for it as soon as possible.

Customer: I think the price is a little expensive. Can you give me a discount?

Customer Service Agent: Dear, I' m sorry to say that our products are small profits but quick turnover. I hope you can understand. If you can place your order today, I will send you a gift and offer free delivery.

Customer: OK, I' ll take it.

Customer Service: Thank you for your arrival.

3. Listen again and tick the answers.

What did the customer service say in the dialogue?

☐ What can I do for you?

☐ Is anybody here?

☐ Can you give me a discount?

☐ I hope you can understand.

☐ I' ll send you a gift and have free postage.

☐ Thank you for your arrival.

4. Pair work.

Follow the above example and make a dialogue.

Tips

①Glad to be of service. 很高兴为您服务。

glad to do sth. 很高兴做……。

②There are enough stock of this commodity, you can book ahead and pay for it as soon as possible. 这个是有货的哦，拍下付款即可。

there be 有…… as soon as possible 尽可能快。

③Can you give me a discount? 可以给我一些优惠吗?

④small profits but quick turnover 薄利多销

place your order 下单

I will send you a gift and offer free delivery. 我会送您一份小礼物及免邮。

⑤Thank you for your arrival. 感谢您的光临。

Dialogue Two

1. Learn the words.

size elastic cheaper delivered pay exchange goods contact

2. Listen and decide whether the sentences are true (T) or false (F).

( ) ①The customer chose large size.

( ) ②The T-shirt can be cheaper.

( ) ③The customer paid with credit card.

( ) ④The goods can be exchanged.

3. Listen again and fill in the blanks.

Customer: Hello!

Customer Service Agent: My dear! I'm here. How can I help you?

Customer: I would like this kind of T-shirt. I'm 1.78 meters tall. What size should I have?

Customer Service Agent: Oh, we suggest that you choose large size. But the T-shirts are ________, you'd better buy a medium size.

Customer: OK. Can it be a little ________________?

Customer Service Agent: Sorry, our T-shirts are of good quality and worth every penny! We offer free shipping on orders today.

Customer: When will this order be ________________ if I buy it?

Customer Service Agent: If you can pay before 4:00 this afternoon today, we will send the ________ off now.

Customer: Can I ________ with cash or credit card?

Customer Service Agent: Sorry, we don't support the payment on the delivery. Credit card, please! You'd better pay by Alipay.

Customer: OK. I will pay by Alipay. But if there is something wrong with the T-shirt, can I ________?

Customer Service Agent: Of course! Don't worry! We offer a guaranty for unconditional refundment within 7 days. When you receive the goods, please check at first. If there is something wrong with the T-shirt, please ________ us at once after taking pictures of it. We will deal with the problem immediately.

Customer: All right. I'll take a medium size.

Customer Service Agent: OK, I'll handle this order right now. Welcome to our shop next time! Goodbye!

Customer: Bye!

Tips

①good quality 优质 worth every penny 物超所值
free shipping on orders 下单免邮
②don't support the payment on the delivery 不支持货到付款
③We offer a guarantee for unconditional refund within 7 days.
我们提供7天无理由退换。

4. Complete the dialogue.

Customer: Hello! ____________________ (有人在吗)?

Customer Service Agent: Hello, dear! Welcome to King's store. ______________, Sir (我有什么可以帮您的吗)?

Customer: I would like this kind of Men's bag. ________________ (还有其他颜色吗) except black?

Customer Service Agent: Of course! There are many colors, such as white, brown, and grey.

Customer: I would like brown one. ______________ (你能给我打个折)?

Customer Service Agent: I'm sorry. The bag is the best seller (最畅销) and today's deals (今天特价). Um, __________________ (给你包邮吧)!

Customer: Can I pay by credit card?

Customer Service Agent: OK.

5. Role play and talk with your partners with the following sentence patterns.

What can I do for you?

Can I help you?

I would like/want to buy...

What size/color do you need?

I pay with/by...

Could you give me some discount...

Can... be a little cheaper?

Free delivery, please!

...

Reading and Writing

1. Look and discuss.

①Do you know what the online customer service is?

②Have you ever been treated badly while being served?

2. Read the information.

Online Customer Service

With the development of Internet, online shopping can be accepted by many people because of its convenience and reasonal price. Online customer service has become a new career, and the demand is also growing.

Online customer service is the service provided online to customers before, during and after a purchase.

As a matter of fact, customer service is a series of activities designed to enhance the level of customer satisfaction, that is, the feeling that a product or service has met the customer's expectation.

As is known to all, not only the goods, but also the service ensures the success of business. Excellent customer service plays an important part in maintaining their customers and attracting new ones.

As a customer service agent, how can you deal with the complaints from the customer? Here are some suggestions:

① Keep calm and listen to the speaker attentively and with great interest. Write down the key points and check with the complainant that each point tells clearly what happened.

② Don' t judge who is right or wrong. Your intention is to find a solution, not to decide who is right or wrong. Ask the complainant for a solution he/she suggests.

③ Follow your organization' s complaint procedures. Don' t make promises you cannot keep. Explain to the complainant why you make such a decision. Show the complainant you take the complaint seriously. If you are wrong, admit it, apologize, and find a way to put things right.

3. Decide true (T) or false (F).

(　　) ①Online customer service is an old job.

(　　) ②Online customer service is the service provided in the shop to customers.

(　　) ③Online customer service only provided the service before a purchase.

(　　) ④Online customer service plays an important role in business.

4. Choose the best answers.

①Online shopping can be accepted by many people because of its ________ and reasonal price.

A. easy　　B. convenience　　C. trouble

②Online customer service is the service provided ________ to customers before, during and after a purchase.

A. online　　B. reception　　C. phone

③________ the goods ________ the service ensures the success of business.

A. Either ; or　　B. Neither; nor　　C. Not only; but also

④If the customer complains, the customer service agent must keep ________ and listen attentively.

A. angry　　B. calm　　C. complaint

5. Choose the activities mentioned in the passage.

a. delivering goods

b. asking for help

c. customer complaint

d. answering the phone

e. attracting new customers

f. keeping calm and listening

g. meeting the customer' s expectation

h. enhancing the level of customer satisfaction

6. **Suppose you are an online customer service agent, what sentences will you use in your service?**

	Sentences
Bargaining(议价)	
Logistics(物流)	
Payment(支付)	
After-sale service(售后)	

Further Study

1. Look and discuss.

Which country does each of the following pictures belong to?

a. Japan b. Rome c. Greece d. Korea e. Egypt

□ □ □ □

2. Find the correct words for the objects in the pictures.

a. pyramids	b. pharaohs	c. mummy	d. sphinx
e. cleopatra	f. Cairo	g. Luxor	h. mosque
i. Red Sea Beach	j. the Nile River	k. Sahara Desert	l. Aswan High Dam

□

□

□

3. Fill in the blanks with the following words in their correct forms.

cloth	tunic	like	Egyptian	straw
linen	barefoot	white	ancient	reach

Unlike most of the people of the ________ Mediterranean, the ________ did not wear just one or two big pieces of ________ wrapped around themselves in various ways. Instead, both men and women in Egypt wore ________ which were sewn to fit them. These tunics were ________ a long T-shirt which ________ to the knees (for men) or to the ankles (for women). They were usually made of ________ and were nearly always ________. Most Egyptians, both men and women, did not seem to have covered their heads with any kind of cloth. They often went ________, but sometimes they wore ________ or leather sandals.

4. Design a T-shirt with the elements of Egyptian culture.

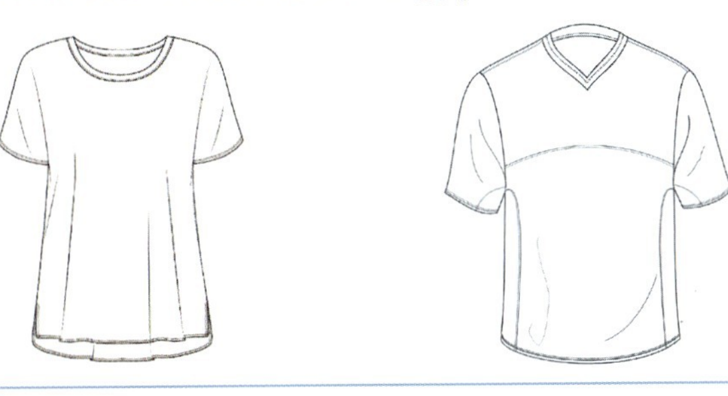

5. Describe your work with a necessary introduction.

Introduction may be as follows:

- Egyptian elements in my mind...
- Color...
- Design (pattern, pyramids...)

Sentence patterns may be as follows:

- My favorite Egyptian element is...
- My design focuses on...
- I use... because its advantage is...

My Progress Check

1. Words I have learned in this unit are:

□ Alipay □ E-commerce □ catalog □ specification
□ service □ stock □ commodity □ profit
□ turnover □ delivery □ elastic □ deliver
□ exchange □ quality □ guaranty □ refundment
□ development □ career □ demand □ purchase
□ enhance □ attract

All together I know ________ words.

More words I know in this unit are:

____

2. Phrases and expressions I have learned in this unit are:

□ credit card □ size chart □ product catalog □ good quality
□ free shipping □ customer service □ place an order □ because of
□ a series of □ deal with □ keep calm □ maintain customer
□ attract customers □ ask for help □ deliver goods □ as a matter of fact

Great! Now I know ________ useful phrases and expressions.

More useful phrases and expressions I know in this unit are:

____

3. I can:

□ identify the detail in product catalog.
□ talk to the customer in the e-commerce.
□ get the general idea of online customer service.

4. I even can:

□ learn how to deal with the customers' complaints.
□ introduce my design with Egyptian elements.

UNIT 8

Window Display

【Goals】

- Identify different elements of window displays
- Learn about fashion accessories
- Learn about the functions of window displays
- Talk to the salesperson about garments in the window display
- Know about a job as a visual merchandiser
- Learn about Italian brands

Warming Up

1. Look and match.

a. bracelet　　b. ear cuff　　c. choker

d. finger bracelet　　e. hair band　　f. ring

g. shoulder bag　　h. canvas bag　　i. nail sticker

□

□

□

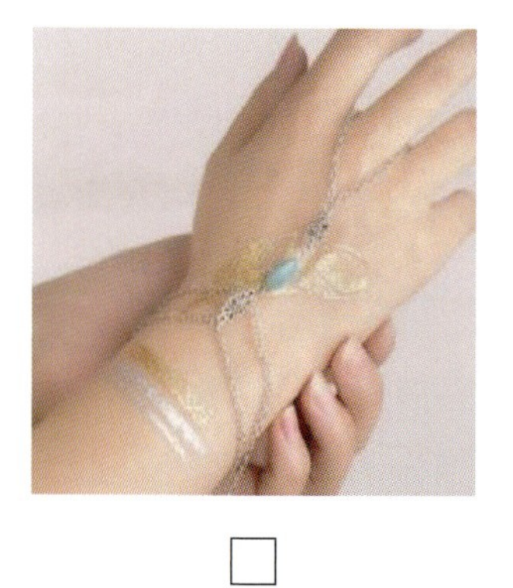
□

□

□

□

□

□

2. Look and guess which seasonal suits they are.

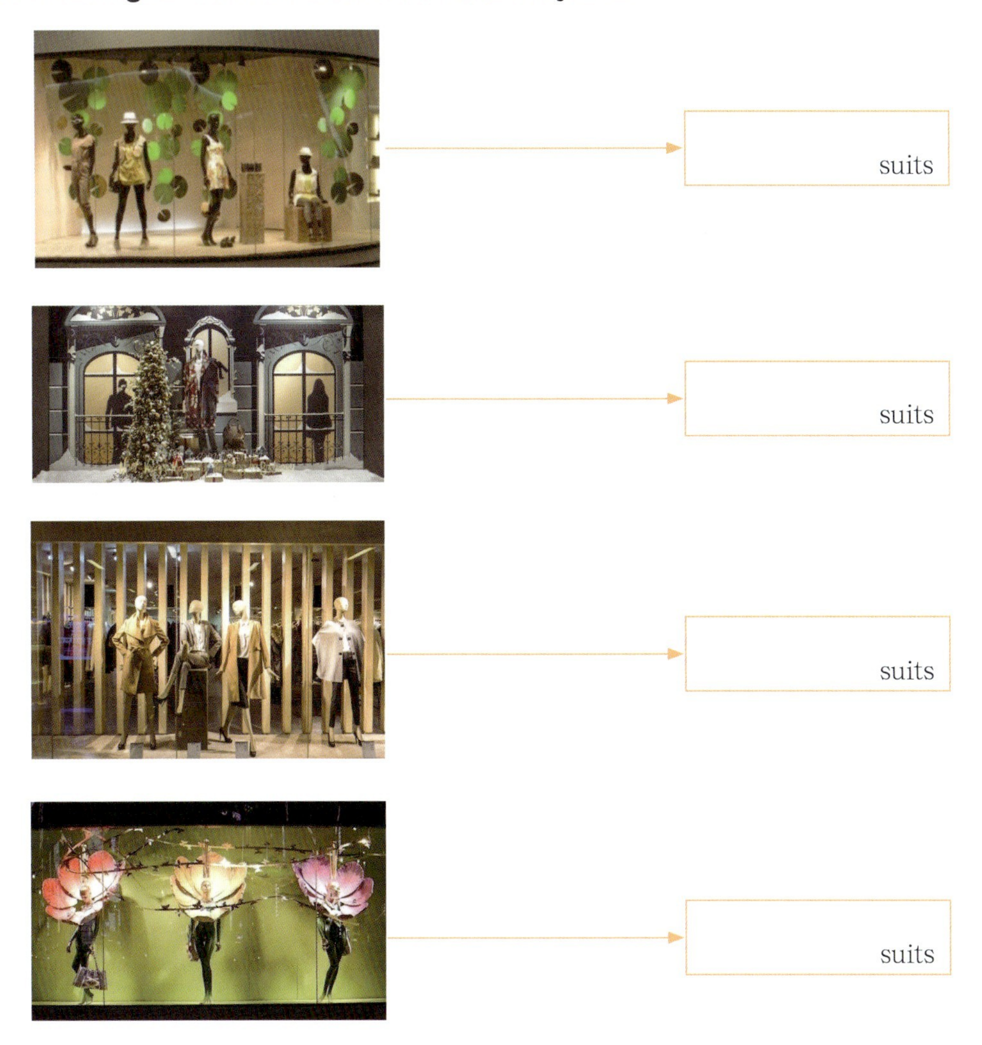

3. Design a theme window display you like.

Listening and Speaking

Dialogue One

1. Learn the words.

visual merchandiser	window display	create	stage	attract
new line	potential buyer	long-term interest		challenging

2. Listen to the dialogue and repeat.

Sally: Hey, Lily! I heard you are looking for a job. What position are you going to apply for?

Lily: Visual Merchandiser(VM). I love window display.

Sally: You mean creating and staging fashion displays?

Lily: You said it!

Sally: It sounds cool. Visual merchandiser plays an important role in a shop.

Lily: Yes. It helps to introduce the new line to potential buyers as well as attracts consumers with goods of long-term interest.

Sally: What an interesting and challenging job! Wish you good luck!

Lily: Sure it is! Thank you.

3. Listen again and tick the answers.

What did we know about a visual merchandiser?

☐ loves window display
☐ creates and stages fashion displays
☐ introduces the new line to potential buyers
☐ attracts consumers with goods of long-term interest
☐ plays an important role in a shop

4. Pair work.

Follow the above example to talk about a job.

Tips

①Visual merchandiser plays an important role in a shop. 陈列专员在店里起着重要的作用。
play an important role in... 在……起着重要作用 (固定搭配)

②It helps to introduce the new line to potential buyers.
它有助于将新季度的产品介绍给潜在顾客。
introduce... to... 把……介绍给……
e.g: Allow me to introduce a friend to you. 请允许我介绍一位朋友给你。

Dialogue Two

1. Learn the words.

collection	striped	pinafore	average	size
fashionable	inspire	expert	wrap	dress

2. Listen and decide whether the sentences are true (T) or false (F).

(　　) ①Maria went shopping in the morning.

(　　) ②Maria tried on one dress.

(　　) ③Maria took two dresses at last.

(　　) ④Window display inspires Maria.

3. Listen again and fill in the blanks.

Maria went shopping one night, and a beautiful window display attracted her, which led her into the shop.

Sales: Good evening! Welcome to Zara!

Maria: Good evening! ________________________________!

Sales: Thank you! ________________________. Would you like to try?

Maria: Sure. I'd like to ________________________ the striped wrap dress, the dummy wears and the white pinafore.

Sales: OK, wait a minute.

(Two minutes later)

Sales: Here are the dresses, ____________________________. And you can try them on in the fitting room.

Maria: Thank you.

(A moment later)

Maria: I love them both. The white one is comfortable, while the striped one looks more fashionable if I put on a black hat as your dummy does. Your window display ______________ a lot!

Sales: I' m glad you like it. Our Visual Merchandiser is an expert in it. Would you like to ______________?

Maria: Yes.

> **Tips**
>
> ①Would you like to try? 需要试一下吗?
>
> ②You can try them on in the fitting room. 您可以在试衣间试穿。

4. Complete the dialogue.

Salesperson: Good morning, ______________ (欢迎来到 Tiffany blue). What can I do for you?

Customer: Your window display is wonderful. ______________ (我想试一下模特穿的那件红色外套).

Salesperson: OK. Wait a moment. ______________ (这是均码的). You can try it on in the fitting room.

Customer: ______________ (这衣服看着很时尚). I like it very much.

Salesperson: Yes, it fits you well. ______________ (就要这一件吗)?

Customer: Yes.

5. Role play and talk with your partners with the following sentence patterns.

I want to try on... the dummy wears.

It feels comfortable.

It looks fashionable.

It is average size.

Would you like to take it?

Reading and Writing

1. Look and discuss.

Can you tell what qualifications a VM should have?

- ☐ High school diploma.
- ☐ Bachelor degree.
- ☐ Merchandising training.
- ☐ Good at designing.
- ☐ Merchandising experience.

2. Read the passage with these questions.

①How many topics can a visual merchandiser aspirant learn about?

②How much can a top VM get a year?

Visual Merchandiser

Are you good at conceptualizing and designing store displays? Have you ever dreamed of working in retail stores or fashion chains? If your answer is yes, then you will have a bright future working as a visual merchandiser.

If you want to pursue a career in visual merchandising, there are plenty of fashion design schools providing excellent course programs. For example, there are Associate Degree Program and Bachelor's Degree Program. Apart from that, visual merchandiser aspirants will also learn about the following topics:

- Store image, layout and concepts
- Lighting fixtures, store displays and signs
- Merchandising methods, plans and theories

- Interior store design ideas
- Concept visualization
- Color theory
- Merchant trend analysis

Employment in the field of visual merchandising is expected to grow faster than the average. In fact, job outlook is quite excellent since more visual merchandising jobs will be generated. To give you an idea on their annual salary estimates, here is a quick rundown of the annual earnings of visual merchandiser professionals according to their work experience.

- Entry Level Visual Merchandiser Professionals: \$21,000 to \$32,000
- Experienced Visual Merchandiser Professionals: \$30,000 to \$45,000
- Top Level Visual Merchandiser Professionals: \$50,000 to \$60,000+

The average yearly earnings of most visual merchandising professionals fall in the \$50,000 income range. Therefore, for those who would like to earn greater income potentials, it would be ideal if you pursue higher educational degrees and acquire more visual merchandising experiences.

3. Decide true (T) or false (F).

(　　) ①If you love designing store displays, you can be a VM.

(　　) ②There are many fashion design schools providing excellent course programs.

(　　) ③VM as a job has a bright future.

(　　) ④Experienced Visual Merchandiser Professionals get no more than \$45,000 per year.

4. Choose the best answers.

①If you want to be a VM, you' d better ________.

A. get a bachelor' s degree　B. work in a fashion store

C. learn some merchandising programs

②Which topic needn' t you learn if you want to be a VM? ________.

A. Color theory　B. Interior store design ideas　C. Building theory

③How much will an entry level VM probably get a year? ________.

A. \$30,000　B. \$40,000　C. \$35,000

④What kind of person wouldn' t be a successful VM? ________.

A. Lack of interest　B. Lack of experience　C. Lack of career

5. According to the passage, plan your future if you want to be a VM.

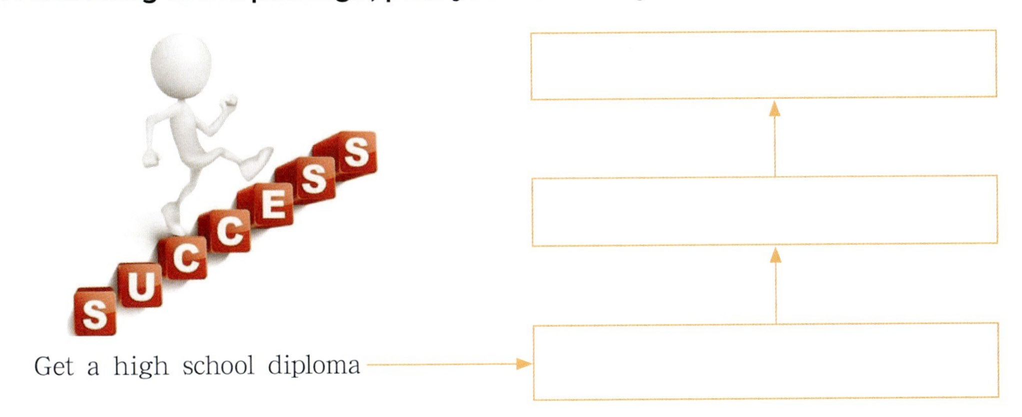

Further Study

1. Look and choose.

Which country does each of the following pictures belong to?

a. Italy	b. Iraq	c. India	d. Russia	e. Spain
f. Vietnam	g. Mongolia	h. South Korea	i. China	

 □

 □

 □

 □

 □

 □

 □

 □

2. Look and write.

Do you know the names of the following garments?

Fill in the blanks.

off collar	suspender skirt	fish tail hem
lantern sleeves	pleated skirt	lace tops

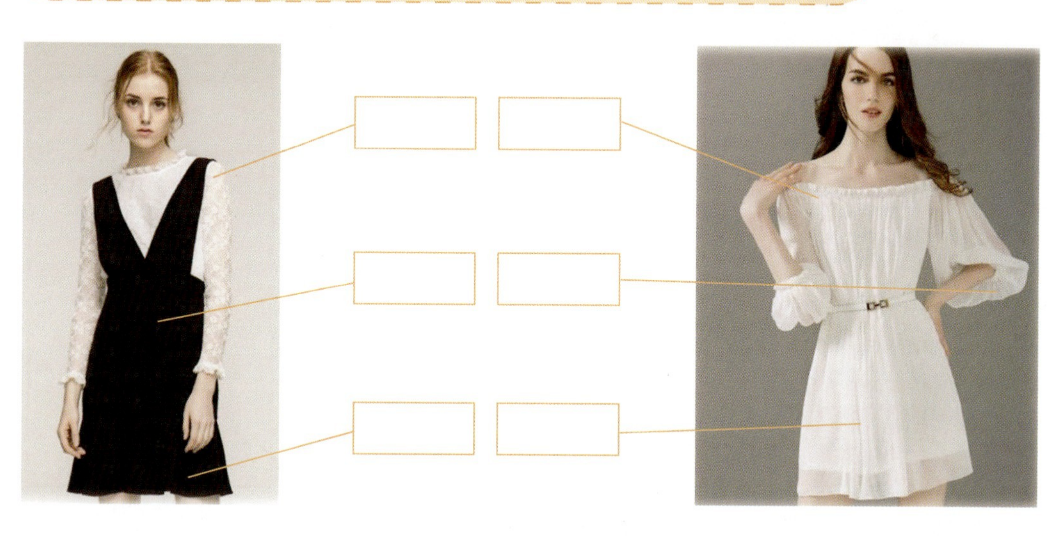

3. Find the correct words for the objects in the pictures.

spaghettis	gladiator shoes	pizza	leaning tower of Pisa	
fashion	espresso	Colosseum	chocolate	Maserati

______________ ______________ ______________

______________ ______________ ______________

______　　______　　______

4. Learn about fashion brands, and pick out the Italian brands.

5. Describe your work with a necessary introduction.

Introduction may be as follows:

- Italian elements in my mind...
- Design (styles, fashion trend, colosseum...)

Sentence patterns may be as follows:

- My favorite Italian element is...
- This is a...
- My design focuses on...
- I use... because its advantage is...

My Progress Check

1. Words I have learned in this unit are:

☐ attract	☐ bracelet	☐ Colosseum	☐ chocolate
☐ challenging	☐ collection	☐ create	☐ choker
☐ expert	☐ espresso	☐ fashionable	☐ fashion
☐ inspire	☐ Maserati	☐ pinafore	☐ pizza
☐ stage	☐ striped	☐ spaghetti	

All together I know ________ words.

More words I know in this unit are:

____

2. Phrases and expressions I have learned in this unit are:

☐ average size	☐ canvas bag	☐ ear ring	☐ ear cuff
☐ finger bracelet	☐ fish tail hem	☐ gladiator shoes	☐ hand-made shoes
☐ hair band Pisa	☐ lantern sleeves	☐ new line	☐ leaning tower of Pisa
☐ lace tops	☐ off collar	☐ potential buyer	☐ shoulder bag
☐ pleated skirt	☐ suspender skirt	☐ window display	☐ wrap dress
☐ visual merchandiser	☐ long-term interest		

Great! Now I know ________ useful phrases and expressions.

More useful phrases and expressions I know in this unit are:

____

3. I can:

☐ talk about fashion accessories.

☐ talk to the sales about garments in the window display.

4. I even can:

☐ describe the job as a VM.

☐ write down some Italian fashion brands.

UNIT 9

In the Theme

【Goals】

- Identify the elements and the features of different styles of clothing
- Know about fashion magazines
- Talk about clothes styles in the theme
- Talk about your favorite style
- Learn to predict the fashion trends
- Work together to design a garment with Thailand elements

Warming Up

1. Look and match.

a. Lady b. National c. Punk
d. Bohemian e. Preppy f. Neutral
g. Casual h. Retro i. Hippies

2. Look and write down the styles.

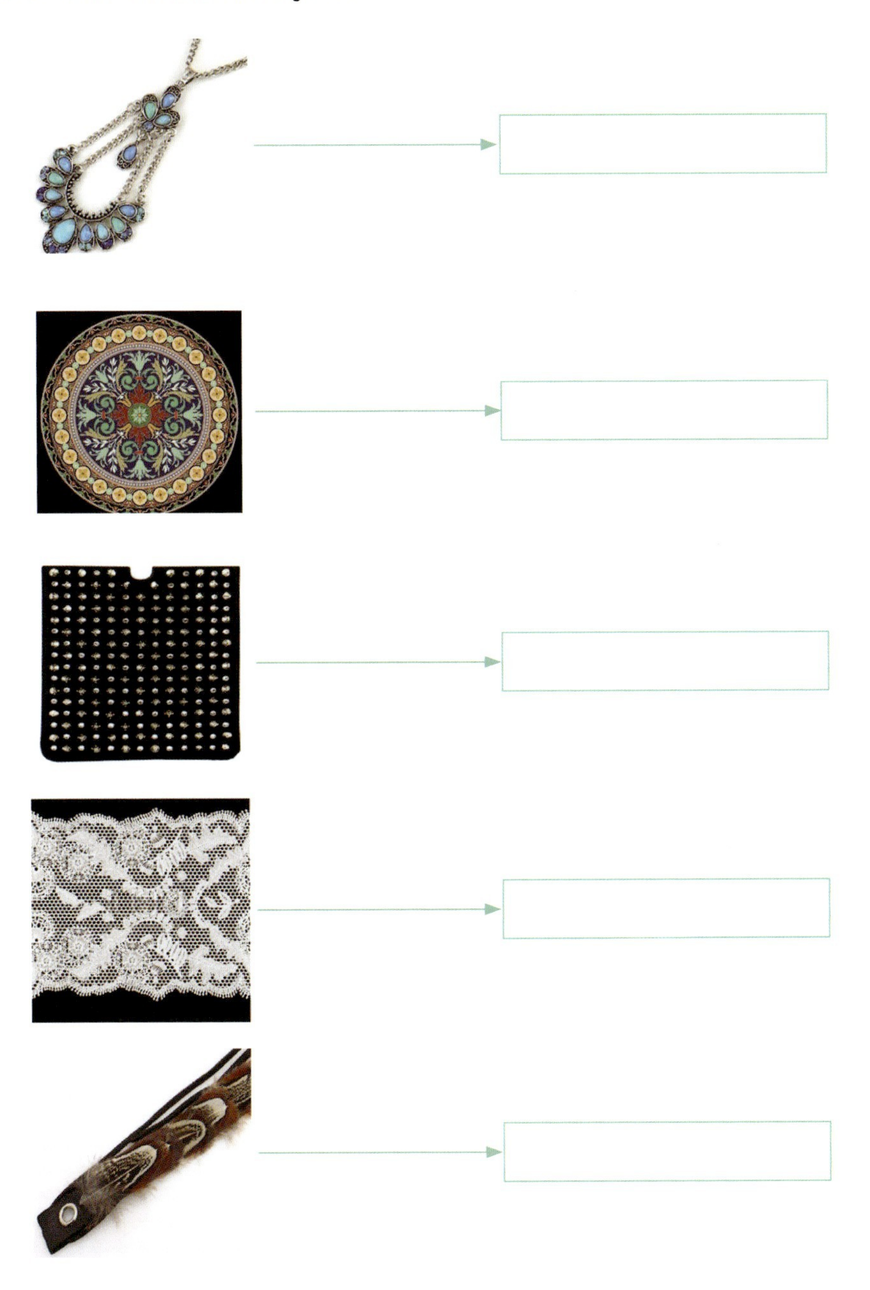

3. Match the features to the styles.

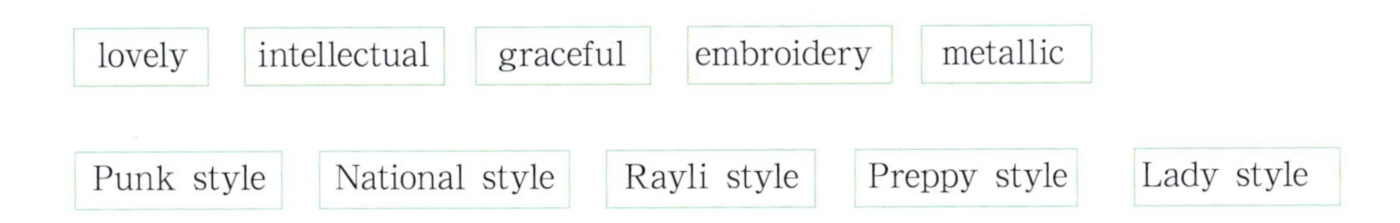

Listening and Speaking

Dialogue One

1. Learn the words.

style graceful suit casual neutral tomboy lace fit

2. Listen to the dialogue and repeat.

Lily: Do you like this style in the magazine?

Mary: It's too graceful. The lady style doesn't suit me.

Lily: What style do you like?

Mary: I like casual wear. I think I dress like Americans.

Lily: Oh, I see. How about the neutral style?

Mary: Yes, I like it too. I look like a Tomboy. How about you?

Lily: I like lace, so I like lady style better.

Mary: Look, this dress fits you well. Do you like this one in the magazine?

Lily: That's what I want. I will take it now.

3. Listen again and tick the answers.

What did Lily say in the theme?

☐ Do you like this style in the magazine?

☐ The lady style doesn't suit me.

☐ What style do you want?

☐ What style do you like?

☐ How about the neutral style?

☐ Do you like this one in the magazine?

4. Pair work.

Follow the above example and make a dialogue.

Tips

① dress like 打扮得(穿得)像……
look like 看起来像……
② fit you well 很适合你

Dialogue Two

1. Learn the words.

dress like　　similar　　casual　　hippies　　way　　surprise　　spend

2. Listen and decide whether the sentences are true (T) or false (F).

(　　) ①Chinese students are sloppier than American students.

(　　) ②Mary likes casual style.

(　　) ③Lucy has been all over China.

(　　) ④Chinese people are too concerned with their clothes.

3. Listen again and fill in the blanks.

Lucy: Do your classmates in China ________ the students here?

Mary: Yes, it's ________. It's even more casual here.

Lucy: What ________ do you like?

Mary: The hippies dressed in very ________ ways. I like it. Have you ever been to ________?

Lucy: No, I haven't.

Mary: If you come to China, you will be ________.

Lucy: Why?

Mary: You will see most of people dressed very ________. They spend a lot of money on clothes now.

Lucy: I'd like to go sometime.

Mary: I think you'll have a good time there.

Tips

①The hippies dressed in very casual ways.　嬉皮士风格都打扮得很随意。

②dress very well　穿得很好

③spend a lot of money on clothes　在衣服上花很多钱

4. Complete the dialogue.

Lisa: What do you think of this coat? ________________ (你认为适合我吗)?

Jay: I don't think so. ________________ (太可爱了).How about this one in ________________ (休闲风格)?

Lisa: It's blue.

Jay: ________________ (有问题吗)?

Lisa: ______________________________ (现在还是流行蓝色的). This style came out last year already.

Jay: ____________ (你不确定) whether the trend of next year the same, right?

Lisa: Yes.

Jay: Oh, ____________ (你最好放弃)and look for some other pieces.

5. Role play and talk with your partners with the following sentence patterns.

What style do you like?

Punk style / Preppy style / Simple style / Lady style...

Do you like...?

How about...?

I like... because...

Reading and Writing

1. Look and match.

□ □ □ □ □

□ □ □ □ □

a. Vogue	b. Cosmopolitan	c. NYLON	d. ELLE
e. L' OFFICIEL W	f. Harper' s Bazaar	g. Marie Claire	
h. Figaro Madame	i. i-D	j. W	

2. Read the passage with these questions.

①Who decided the new fashion? Why?

②How did the life-style changes change people's buying habits?

Trend Prediction

Today fashion begins and ends with the consumers. It is the consumer, not the designer or manufacturer, who determines what will be in fashion by accepting one style and rejecting another. Price and sales promotions do not dictate what styles will be accepted by consumers. These can only encourage or retard the process of new fashion.

Consumers must be considered in the context of outside influences, such as social influences, cultural and political influences, economic influences, technological influences, and so on. If we learn to observe these influences we can better foresee fashion development.

Changes in people's attitudes and life-styles change their fashion and buying habits. People want fashion appropriate to their interests and activities. For better education and exposed to newer ideas, women want a wider fashion choice. To participate in sports, women need more active clothing. For jobs they need practical and businesslike clothes, such as suits. The shortening work week and a more casual life-style have increased men's participation in home life and leisure activities. Their wardrobes formerly limited to suits, slacks and sports shirts have expanded along with their activities, and increased choices have made them more fashion conscious.

3. Match the English words with the Chinese meanings.

a. consumer　b. designer　c. manufacturer　d. determine　e. reject

f. dictate　g. retard　h. foresee　i. appropriate　k. wider

命令（　）　预见（　）　设计师（　）　消费者（　）　更宽的（　）

做决定（　）　阻止（　）　适当的（　）　制造商（　）　拒绝（　）

4. Decide true (T) or false (F).

(　) ①Fashion is decided by manufacturer.

(　) ②Price and sales promotions can influence the styles.

(　) ③Sales promotions can better foresee fashion development.

(　) ④People want fashion appropriate to their interests and activities.

(　) ⑤For jobs women need practical and businesslike clothes.

5. Choose the best answers.

①Who can accept one style and reject another? ________.

A. Designer　B. Consumer　C. Manufacturer

②________can not encourage the process of new fashion.

A. Price　B. Sales promotion　C. Both A and B

③To participate in sports, women need ________.

A. sports wear　B. suits　C. dress

④________ have increased men's participation in home life and leisure activities.

A. The shortening work week and a more casual life-style

B. Salary increase and a more casual life-style

C. Salary increase and healthy keeping

⑤Men's wardrobes have ________ along with their activities, and increased choices.

A. expanded　B. diminished　C. changed

6. Answer the questions.

①What's the writer's opinion on predicting the fashion trend?

____

②What influences will be considered by the consumers?

____

③What's the benefit of observing the influences?

____

Further Study

1. Look and discuss.

Which country does each of the following pictures belong to?

a. Thailand　　b. India　　c. England　　d. America

□　□　□　□

2. Look and choose.

a. Phuket Island	b. Grand Palace	c. Boxing champion
d. Wat Arun	e. Chiengmai	f. museum
g. Chulalongkorn University	h. Wat Pho	i. Marble Temple
j. Jade Buddha Temple	k. Pattaya	l. Ssangyong temple

□

□

□

□

□

□

□

□

□

□

□

□

3. Find the correct words for the objects in the pictures.

lady-boy	elephant safari	seafood
durian	floating market	Plowing Day
Ghost festival	beach	fried crab with curry sauce

____

____

____

____

____

____

4. Design one of the garments with the elements of Thailand culture you like.

5. Describe your work with a necessary introduction.

Introduction may be as follows:

- Thailand elements in my mind...
- Type of styles...
- Fashion trend...
- Design (style, fashion trend, temple...)

Sentence patterns may be as follows:

- My favorite Thailand element is...
- This is a...
- My design focuses on...
- I use... because its advantage is...

My Progress Check

1. Words I have learned in this unit are:

□ neutral	□ punk	□ casual	□ graceful
□ hippies	□ lace	□ consumer	□ manufacturer
□ reject	□ promotion	□ rayli	□ political
□ foresee	□ expose	□ attitude	□ wider
□ leisure	□ slack	□ expand	□ style
□ spend	□ suit	□ fit	□ retro

All together I know ________ words.

More words I know in this unit are:

____

2. Phrases and expressions I have learned in this unit are:

□ dress like	□ fashion trend	□ casual wear	□ neutral style
□ fits you well	□ look like	□ concerned with	□ spend on
□ casual way	□ give up	□ look for	□ life-style
□ buying habit	□ sales promotion	□ political influences	□ participated in

Great! Now I know ________ useful phrases and expressions.

More useful phrases and expressions I know in this unit are:

____

3. I can:

□ identify different styles of clothing.

□ identify the elements and the features of different styles.

□ know about fashion magazines.

□ talk to my friends about clothes styles in the theme.

□ talk to my friends about my favorite style.

4. I even can:

□ learn to predict the fashion trends.

□ work together to design a garment with Thailand elements.

UNIT 10

In the Studio

【Goals】

- Know something about the studio
- Learn about works of the garments
- Learn how to make a resume
- Know how to talk to the manager about your job intension
- Learn to recommend yourself
- Learn about some world brands

Warming Up

1. Look and match.

a. drawing designs
b. dress designer
c. clothing sales
d. garment studio
e. the designer of apparel cutting
f. garment pattern maker
g. having interview
h. sewing
i. doing computer graphics

□

□

□

□

□

□

□

□

□

2. Look and guess what they are doing.

Listening and Speaking

Dialogue One

1. Learn the words.

Personality	approach	enthusiastically	extremely	capable
subjectivity	avantgarde	certification	cooperative	

2. Listen to the dialogue and repeat.

Manager: Tell me a little bit about yourself.

Applicant: My name is Jim and I live in Guangzhou. I was born in 1990. My major was dress designing.

Manager: What kind of personalities do you think you have?

Applicant: Well, I approach things very enthusiastically, I think. I don't like to leave things half-done. I'm very organized and extremely capable.

Manager: What are your weaknesses and strengths?

Applicant: Well, I'm afraid I'm a poor speaker, so I've been studying how to speak in public. But I'm good at understanding the costume's subjectivity and effectivity.

Manager: Do you have any licenses or certifications?

Applicant: I have a garment designer certification.

Manager: What software do you use for fashion design?

Applicant: I'm good at using Photoshop 7.0, CorelDRAW, AuotCAD, Dreamweaver MX 2004, Flash MX 2004, Fireworks MX 2004 and so on.

Manager: Which styles do you like best?

Applicant: I like the sports style and avantgarde style.

Manager: How do you relate to others?

Applicant: I'm very cooperative and hard-working. I also have good teamwork spirits.

Manager: How can I contact you when we reach our decision?

Applicant: You can call me at this number at any time.

Manager: OK, see you.

Applicant: Thank you, see you.

3. Listen again and tick the answers.

What personality does the applicant have?

☐ He approaches things very enthusiastically.

☐ He is creative and honest.

☐ His major is dress designing.

☐ He is very organized and extremely capable.

☐ Never leave things half-done.

4. Pair work.

Follow the above example to talk about a job.

①Tell me a little bit about yourself. 请介绍一下你自己。

②My major was dress designing. 我的专业是服装设计。

③What are your weaknesses and strengths? 你的弱点和优点是什么？

Dialogue Two

1. Learn the words.

combined with	internship	technical	qualification
deliver	theoretical aspects	housing packages	insurance

2. Listen and decide whether the sentences are true (T) or false (F).

(　　) ①Clare is coming for an interview as required.

(　　) ②The company is global.

(　　) ③Clare's major is suitable for this position.

(　　) ④At last, Clare didn't get the job.

3. Listen again and fill in the blanks.

(Clare is having an interview in a company.)

Manager: Good morning, Mr. Li! Please sit down!

Clare: Good morning, sir. Thank you. I'm coming for ________ as required.

Manager: Fine, thank you for coming. Why do you ________ the position in our company?

Clare: Well, sir, your company is global, so I feel I can gain more by working in this field. What's more, my college training combined with my internship should qualify me for this particular job. I am sure I will be successful. ________ is dress designing that is suitable for this position.

Manager: Er, I see. Among the qualities we ____________ is technical abilities and practical skills. A lot of university graduates have paper qualifications. But can they deliver? What do you think?

Clare: Yes, well, at my college, we have been working mainly on theoretical aspects so we have a firm foundation. But each semester we have been asked to work on these practical projects.

Manager: How do you know about this company?

Clare: Your company is very famous. I heard of much praise for your company.

Manager: Do you have any particular conditions that you would like the company to __________ consideration?

Clare: Yes, do you have ________________________ a housing packages, medical insurance, unemployment insurance and so on?

Manager: Yes, we do. If you are hired, when will it be convenient for you to begin to work?

Clare: Tomorrow is OK, if you like.

Tips

①Why do you apply for the position in our company?
你为什么要申请我们公司的这个职位?

②My major is dress designing that is suitable for this position.
我的专业就是服装设计，我认为我很适合这个职位。

4. Complete the dialogue.

Manager: Good morning, ______________________. (自我介绍一下吧).

Applicant: Good morning, sir. My name is Jim and I was born in 1990.

______________________________________.(我的专业是服装设计。)

Manager: OK. ___________________________? (你的强项和弱项是什么?)

Applicant: I'm good at understanding the Costume's Subjectivity and Emotion.

Manager: Well. ___________________________?(你为什么要申请来我们公司?)

Applicant: Because the management style of your company is humanistic and open minded, so I feel I can gain the most from working in this kind of environment.

Manager: Yes, we do. If you are hired, ____________________? (你什么时候方便来上班?)

Applicant: Tomorrow is OK, if you like.

5. Role play and talk with your partners with the following sentence patterns.

Tell me a little bit about yourself.

I'm suitable for this position.

What are your weaknesses and strengths?

Why do you apply for the position?

How do you know about the company?

Reading and Writing

1. Look and discuss.

Can you tell what qualifications a fashion designer should have?

☐ He has a bachelor's degree.

☐ He is cooperative and hard-working.

☐ He has technical ability and practical skills.

☐ He is good at designing.

☐ He has designing experiences.

2. Read the passage with these questions.

①What information should be included in the Resume?

②How can you recommend yourself in the Resume successfully?

Resume

Personal Information:

Name: Wang Ming

Health: Excellent

Gender (Sex): Female **Age:** 20

Place of Birth: Guangzhou, Guangdong, China **Marital Status:** Single

Address: No.18 Zhongshan Road, Guangzhou **Zip Code:** 510000

Tel: (020) 2368-7098 **Mobile Phone:** 1382-888-6666

E-mail Address: Wangming666@163.com

Job intention: Fashion designer; garment pattern maker; merchandiser

Education: September, 2012—July, 2015
Zhenjiang Secondary Vocational School

Major: Certificate of Technical Secondary School education;
Garment Designer Certificate

Part-time Work Experience: December, 2014—May, 2015
work as fashion designer's assistant of GOELIA

Language Proficiency

English: Accurate & Fluent LCCI: Excellent

Computer Proficiency

Pass the National Computer Rank Test; familiar with Photoshop 7.0, CorelDRAW, AuotCAD, Dreamweaver MX 2004,Flash MX 2004, Fireworks MX 2004, and so on.

Other Specialities

Knowledge of commerce and foreign trade; useing hand-painting and computer designing skillfully; good at pattern making and apparel draping.

Self-judgment

Honest, energetic, fashion minded, having strong ambition and determination to succeed; pleasant personality with initiative and drive; prepared to work hard, ability to learn; creative while possessing a great team spirit.

3. Decide true (T) or false (F).

( ) ①Wang Ming wants to be a sales.

( ) ②There is much software that can be used to design clothes.

( ) ③Wang Ming isn't good at English.

( ) ④Wang Ming is good at hand-painting and computer designing.

4. Choose the best answers.

①If you want to be a fashion designer, you' d better ________.

A. have a garment designer certificate

B. work in a fashion store　　C. learn some English

②Which topic needn' t you learn if you want to be a fashion designer? ________.

A. Color theory　　B. Pattern making and apparel draping

C. Building theory

③Which software may probably be used in fashion designing? ________.

A. CorelDRAW　　B. Office 2000　　C. Internet

5. Design your own resume.

Further Study

1. Look and choose.

Which country does each of the following pictures belong to?

a. Philippine	b. Brazil	c. Germany	d. Netherlands	
e. Australia	f. France	g. Singapore	h. Italy	i. India

□

□

□

□

□

□

□

□

□

2. Look and write.

Do you know what jobs the clothing graduates can do?

Fill in the blanks with the following words.

freehand sketching	trainer	computer graphics
draping	apparel pattern making	drafting

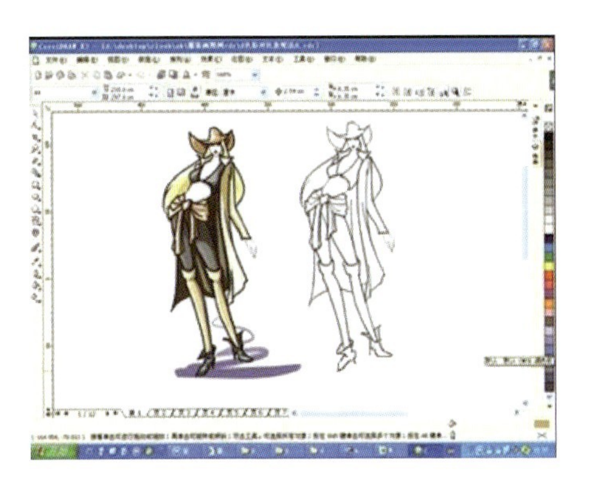

____

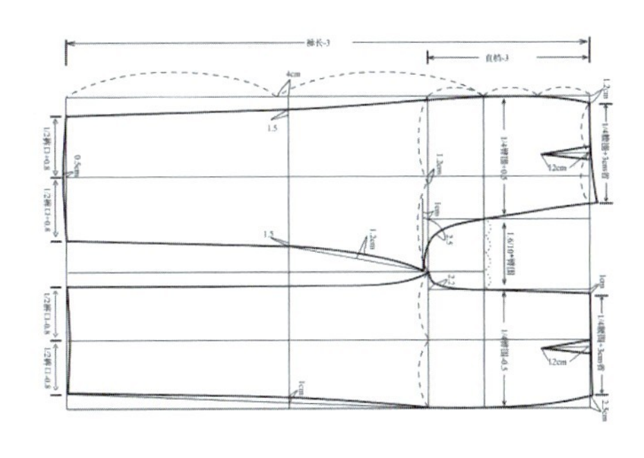

____

____

____

____

____

3. Find the correct words for the objects in the pictures.

hand sewing	machine sewing	ironing and pressing
apparel pattern	the certification of costume designer	
sign of zipper	washing symbols	overedging

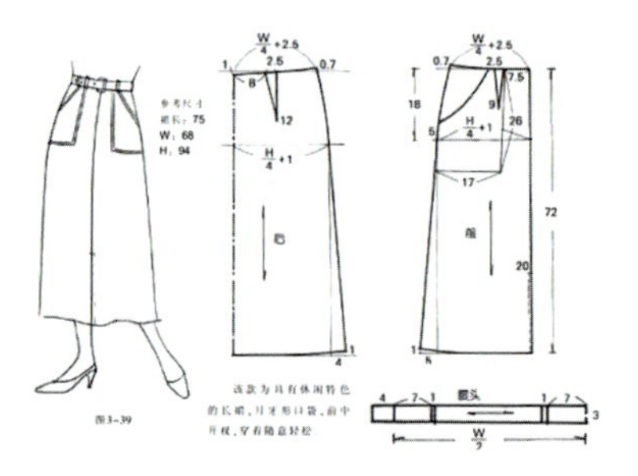

____

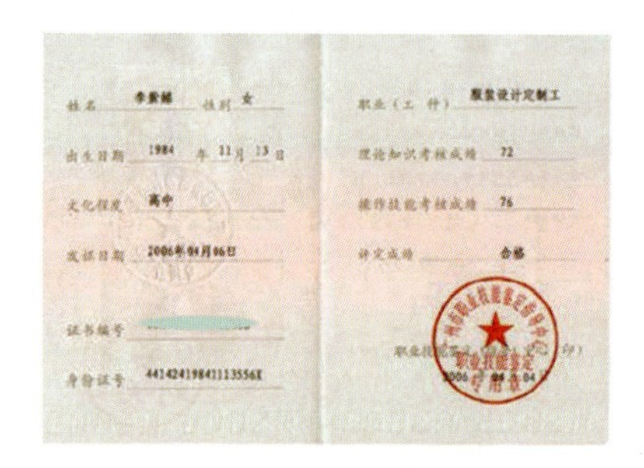

____

服装洗涤标志中英对照说明

	可以手洗 WASHING WITH HAND
	不可机洗 DO NOT MCHINE WASH
P	干洗 DRY CLEAN ONLY
	不可漂白 DO NOT BLEACH
	蒸汽熨烫 STEAM PRESSING
	不可转笼翻转干燥 DO NOT TUMBLE DRY

____

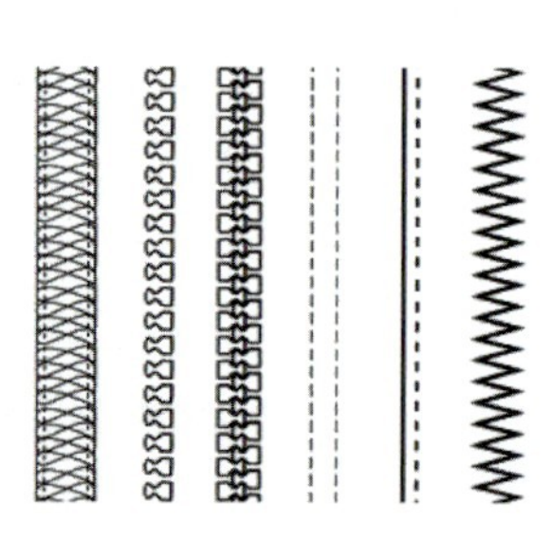

____

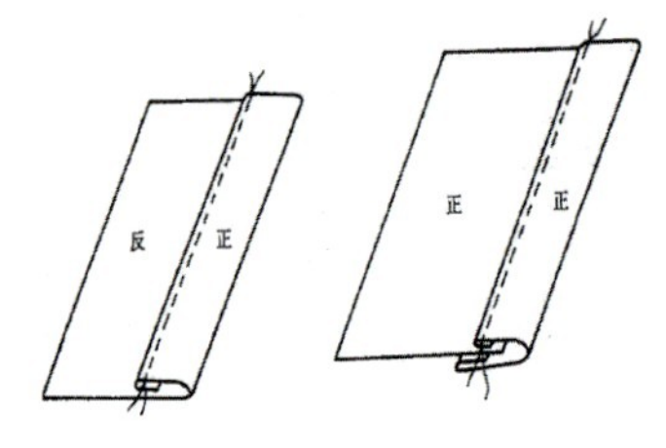

____

名称	擦针	线丁	擦针
符号			
名称	纳针	倒勾针	拱针
符号			
名称	三角针	杨树花针	线袢
符号			
名称	打套结	锁眼	钉扣
符号			

____

名称	明线	双止口明线	碎褶	折裥	明裥
符号					
名称	暗裥	省	开省号	塔克线	司马克
符号					
名称	罗纹	橡筋	扣眼位	扣位	眼刀
符号					

____

名称	烫干	烫圆	拉烫	缩烫	归烫	拔烫
符号	90°C	120°C	120°C	140°C	140°C	140°C
名称	湿烫	干烫	盖布烫	不能烫	粘合烫	蒸汽烫
符号	300°C	100°C	500°C	0°C	200°C	500°C
说明	符号中的数字表示熨烫温度，温度的高低应根据面料测试的承受度数来标注					

____

4. Learn about fashion brands and classify brands by countries.

France	America	Philippine

My Progress Check

1. Words I have learned in this unit are:

- ☐ apparel
- ☐ personality
- ☐ approach
- ☐ enthusiastically
- ☐ extremely
- ☐ capable
- ☐ subjectivity
- ☐ certificates
- ☐ overedging
- ☐ co-operative
- ☐ draping

All together I know ________ words.

More words I know in this unit are:

____

2. Phrases and expressions I have learned in this unit are:

- ☐ capable subjectivity
- ☐ avant-garde style
- ☐ freehand sketching
- ☐ computer graphics
- ☐ apparel pattern
- ☐ making hand sewing
- ☐ machine sewing
- ☐ ironing and pressing
- ☐ business card
- ☐ sign of zipper
- ☐ sign of coincide
- ☐ avant-garde style
- ☐ hand sewing
- ☐ washing symbol
- ☐ the certification of costume designer

Great! Now I know ________ useful phrases and expressions.

More useful phrases and expressions I know in this unit are:

____

3. I can:

- ☐ talk about my future job.
- ☐ make a resume.
- ☐ describe jobs with proper tense.

4. I even can:

- ☐ read employment ads.
- ☐ write a cover letter to apply for a job.

UNIT 11

In the Fashion Show

【Goals】

- Identify famous fashion designers
- Identify famous fashion brands
- Talk to the staff about holding a fashion show
- Talk to the customers with sales promotion
- Understand classic look in clothing
- Work together to design a garment with Korean elements

Warming Up

1. Look and match.

a. Giorgio Armani
b. Issey Miyake
c. Louis Vuitton
d. Gabrielle Bonheur Chanel
e. Valentino Garavani
f. Gianni Versace
g. Mario Prada
h. Guccio Gucci
i. Christian Dior

☐

☐

☐

☐

☐

☐

☐

☐

☐

2. Match the cities with the following pictures.

Paris of France | Tokyo of Japan
New York of the United States of America | Milan of Italy
London of the United Kingdom

____

____

____

____

____

3. Classify brands by cities and put the above brands into the form.

Cities	Brands

Listening and Speaking

Dialogue One

1. Learn the words.

topic   concise   runway   special   creative   hesitate   add up

2. Listen to the dialogue and repeat.

Lucy: The topic for this fashion collection would be classic. A large scale of black, white collection, together with bright style would be the main tone.

Jack: I see. So it would be concise for the stage.

Lucy: What about style for the runway? Any good ideas?

Jack: It can be a little bit special. For example, it can be an S shape. And the scene and models inside that should respond to our theme.

Lucy: It sounds creative.

Jack: How many styles need to be shown?

Lucy: 60 styles.

Jack: OK, 15 models would be enough.

Lucy: Well, it is up to you. If too few, there will be not enough time for clothes changing. If too many, it will add up the cost.

Jack: No problem.

3. Listen again and tick the answers.

What did the staffs discuss in the fashion show stage?

☐ How many styles need to be shown?

☐ It is up to you.

☐ What style do you want?

☐ What about the topic?

☐ What about style for the runway?

☐ It sounds creative.

4. Pair work.

Follow the above example and make a dialogue.

Tips

①If too few, there will be not enough time for clothes changing.
如果(模特)太少，换衣服的时间就不够。
②If too many, it will add up the cost.
如果(模特)太多，就会增加成本。
add up 等于

Dialogue Two

1. Learn the words.

handle	hesitated	promotion	concern
official account	cash	credit card	password

2. Listen and decide whether the sentences are true (T) or false (F).

(　　) ①I love the suit which we have seen in the fashion show on TV.
(　　) ②I feel a little hesitated about the price.
(　　) ③We don't have promotions.
(　　) ④If you follow our WeChat official account, you can get 40% off for our garments.

3. Listen again and fill in the blanks.

Jenny: That is what I want, the suit that can handle my next week's job ________. I love the suit which we have seen in the _______ show on TV. What do you think?

Emma: Well, maybe you can try to put them on. Excuse me, please get the suit on the fashion show for size M to my friend.

Salesclerk: OK.

(After a few minutes)

Emma: You look so pretty. Do you like them?

Jenny: I like the ________, but I feel a little hesitated about the price.

Salesclerk: Don't worry, we have promotions. If you follow our WeChat ________ ________, you can get 40% off for our ________.

Emma: Oh, it is not expensive, right?

Jenny: Great, I would like the suit. That's all.

Salesclerk: Cash or ________________________________, please?

Jenny: Credit card.

Salesclerk: OK. Enter the password and sign, please. Welcome again.

Tips

That is what I want, the suit that can handle my next week's job interview.
这正是我想要的套装，它能解决我下周工作面试的着装问题。
If you follow our WeChat officical account, you can get 40% off for our garments.
如果您关注我们的微信公众号，你就可以得到六折优惠。

4. Complete the dialogue.

Daisy: ________________________(下周我要去面试工作). Would you please go shopping with me?

Jasmin: Of course. ________________________(看，电视里时装秀的那套套装).

Daisy: ________________________(那是我想要的套装).

Jasmin: OK. Do you like them?

Daisy: ________________________(我喜欢这个类型，但是价格方面我有点迟疑). What do you think?

Jasmin: Yes, ________________________(如果他们有促销活动就好了).

5. Role play and talk with your partners with the following sentence patterns.

The topic for this fashion collection would be...

It can be a little bit...

That is what I want...

If...

Reading and Writing

1. Look and discuss.

Which one is classic?

A

B

C

D

2. Read the passage with these questions.

①Do you have any classic elements in your clothes?

②What are they?

Classic Look

A classic means something not easy to be dated. Anytime you put it on, it does not look out-of-date; it looks elegant, chic and stylish.

Some styles or designs continue to be considered in good taste over a long period of time. They are always on the usual movement of styles through the fashion life cycle. A classic is a style or a design that satisfies a basic and remains in general fashion acceptance for an extended period of time.

Depending upon the fashion statement one wishes to make, a person may have only a few classics or may have a wardrobe of mostly classics. A classic is characterized by simplicity of design which keeps it from being easily dated. The Chanel suit is an outstanding example of a classic. The simple lines of the Chanel suit have made it acceptable for many decades, and it reappears now and then as a fashion. Many women always have a Chanel suit in their wardrobes. Other examples of classics are blue denim jeans, blazer jackets, cardigan or turtleneck sweaters, and button-down oxford shirts.

Among accessories, the shoes, the loafer, the one-button gloves, the pearl necklace, and the clutch handbag are also classics. For young children, overalls and one-piece pajamas have become classics.

3. Decide true (T) or false (F).

(　　) ①Every time you put the classic on, it looks bad.

(　　) ②Classics are out of the usual movement of styles through the fashion life cycle.

(　　) ③A person may not have a few classics or may not have a wardrobe of mostly classics.

(　　) ④Young children have overalls and one-piece pajamas for classics.

4. Choose the best answers.

①What does a classic look? ________.

A. Ugly　　B. Elegant　　C. Old

②Some styles or designs continue to be considered ________ over _______ of time.

A. in good taste; a short period　　B. in bad taste; a long period

C. in good taste; a long period

③The Chanel suit is ________ example of a classic.

A. an excellent　　B. an ordinary　　C. an outstanding

④________ have a Chanel suit in their wardrobes.

A. Men　　B. Women　　C. Men & Women

5. Match the words with their definitions.

denim	A sweater with a short round collar that fits closely around your neck.
men's suit	A knitted woolen sweater that you can fasten at the front with buttons or a zip.
cardigan	A kind of jacket which is often worn by members of a particular group, especially schoolchildren and members of a sports team.
turtleneck	A men's suit consists of a jacket, trousers, and sometimes a waistcoat, all made from the same fabric.
blazer jacket	(usually plural) Close-fitting trousers of heavy denim for manual work or casual wear.

6. Answer the following questions.

①What's the classic in your opinion?

____

②What classic clothes do you have?

____

③What classic clothes would you like to have in your wardrobe and why?

____

Further Study

1. Look and discuss.

Which country does each of the following pictures belong to?

Democratic People's Republic of Korea　　　　Republic of Korea

______________________　　　　______________________

2. Look and choose.

a. an upper garment	b. shoes	c. a skirt
d. sleeves	e. shoulder straps	f. top of the skirt

3. Find the correct words for the objects in the pictures.

East Gate	hanbok	traditional house	kimchi	taekwondo
Jeju Island	Lotte World	ginseng chicken soup		Teddy bear museum

________________ ________________ ________________

________________ ________________ ________________

________________ ________________ ________________

4. Design one of the garments with the elements of Korean culture.

5. Describe your work with a necessary introduction.

Introduction may be as follows:

- Korean elements in my mind...
- Type of garments...
- Color and tones...
- Design (color, tones, types...)

Sentence patterns may be as follows:

- My favorite Korean element is...
- This is a...
- My design focuses on...
- I use... because its advantage is...

My Progress Check

1. Words I have learned in this unit are:

☐ topic	☐ handle	☐ elegant	☐ wardrobe
☐ concise	☐ promotion	☐ chic	☐ simplicity
☐ runway	☐ concern	☐ stylish	☐ decade
☐ special	☐ cash	☐ satisfy	☐ blazer
☐ creative	☐ password	☐ acceptance	☐ cardigan
☐ hesitate	☐ classic	☐ extend	☐ turtleneck

All together I know ________ words.

More words I know in this unit are:

____

2. Phrases and expressions I have learned in this unit are:

☐ official account	☐ add up	☐ one-piece
☐ not enough	☐ men's suit	☐ the United States of America
☐ button-down	☐ the United Kingdom	☐ traditional house
☐ pump-style	☐ one-button	☐ shoulder straps

Great! Now I know ________ useful phrases and expressions.

More useful phrases and expressions I know in this unit are:

____

3. I Can:

☐ identify famous fashion designers.

☐ identify famous fashion brands.

☐ talk to the staff about holding a fashion show.

☐ talk to the customers about sales promotion.

4. I even can:

☐ learn to know a classic look.

☐ work together to design a garment with Korean elements.

UNIT 12

In the Recycling Industry

【Goals】

- Identify the procedure of clothing recycling
- Identify clothing recycling containers
- Talk to a friend about recycling
- Talk to family members about reuse
- Understand textile recycling
- Work together to design a garment with Russian elements

Warming Up

1. Look and match.

a. donation to hope primary schools
b. sorting
c. 70% rags or other recovered materials
d. a recycling company
e. clothing recycling container
f. 500 tons waste textiles
g. 15% donation to export to other countries
h. processing
i. used clothing

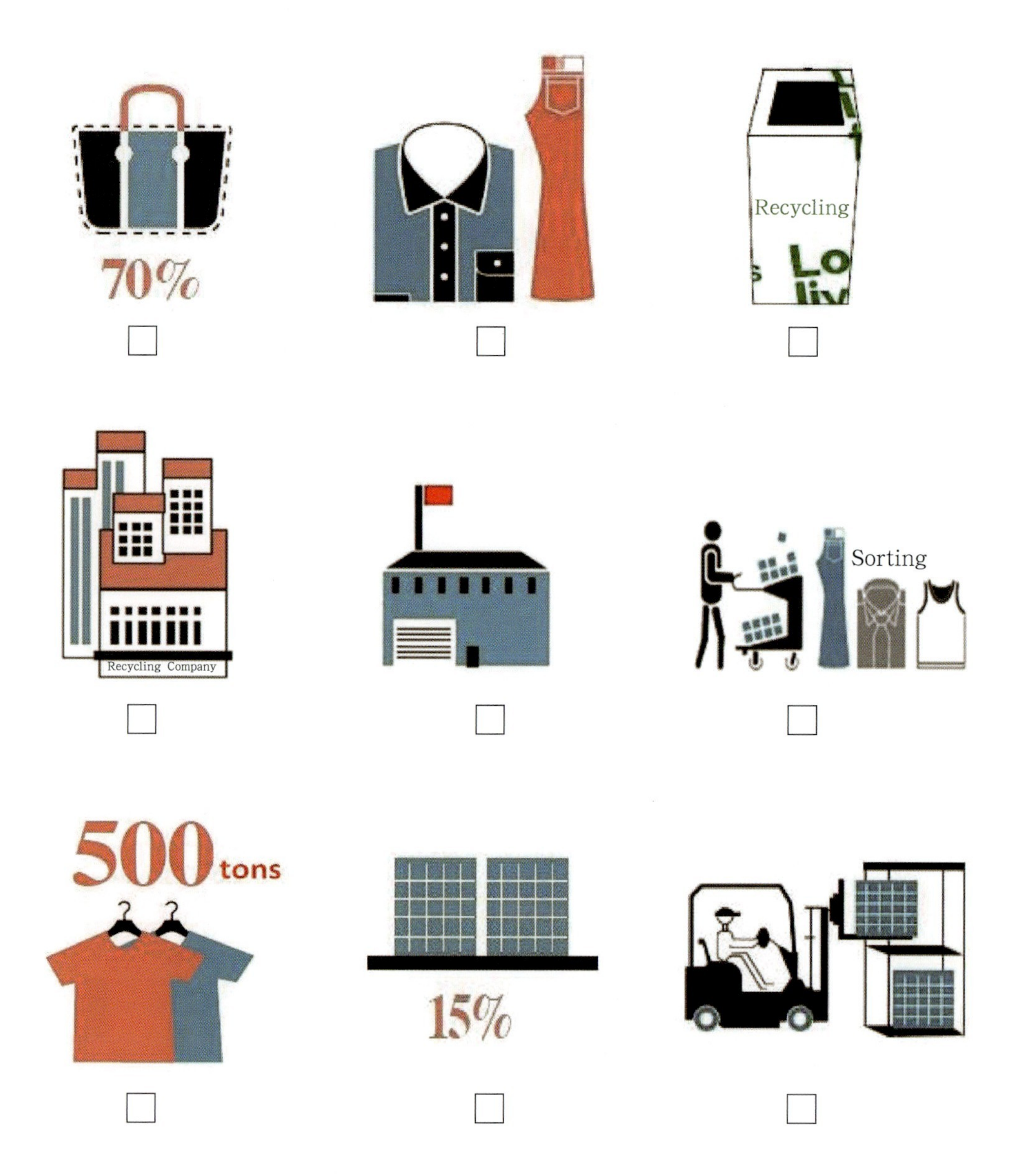

2. Look and guess.

①The name of the organization:

____

②The slogan of the organization:

____

③The amount of the clothing:

____

④The country of the citizen:

____

⑤The period of the time:

____

⑥The item of the waste:

____

⑦The unit of the weight:

____

⑧The way of the waste:

____

3. Put the items to the correct clothing recycling containers.

jeans	sandals	a skirt	a shirt
trousers	stocking	sneakers	boots
mosquito net	high heel shoes		

Clothes

Shoes

Listening and Speaking

Dialogue One

1. Learn the words.

can	enlighten	soccer shirt	quilt
coziness	python	sort	recycling center

2. Listen to the dialogue and repeat.

Candy: Do you know something about recycling? Recycling isn't just about cans, glass and newspapers.

Johnson: Really, please enlighten us. Oh, wise one! We would like to know more!

Candy: Well, "recycling" means using old things to make new things. For example, remember when Granny cut up your old soccer shirts and made them into a quilt? That's recycling!

Johnson: I'm not so sure about the coziness of that particular item, though the cat seemed to enjoy it. Hey, the cat recycled the quilt into a cat bed! That's recycling, right?

Candy: Sure. *Pets Who Recycle* sounds like a reality TV show. "This week: *Cats with Cans and Pythons with Paper*." Clearly, we can do a lot more than just sort bottles, cans, clothes and paper to send to the recycling center.

3. Listen again and tick the answers.

What did the friends say in the conversation?

- ☐ That's recycling, right?
- ☐ *Pets Who Recycle* sounds like a reality TV show.
- ☐ Hey, the cat recycled the quilt into a cat bed!
- ☐ Please don't enlighten us.
- ☐ Do you know something about recycling?
- ☐ Recycling means using old things to make new things.

4. Pair work.

Follow the above example and make a dialogue.

Tips

I'm not so sure about the coziness of that particular item, though the cat seemed to enjoy it.

我不太确定这个物品是否舒适，但是那只猫看起来很喜欢它。

I'm not so sure about... 我不太确定……

though 即使；虽然，尽管；纵然

Dialogue Two

1. Learn the words.

ballet shoes　grow　plant　juicy　reusing　jam jars　storage　locally

2. Listen and decide whether the sentences are true (T) or false (F).

(　　) ①She found the shoes at the beginning.

(　　) ②The mother grew potatoes in shoes.

(　　) ③She can't accept the way of reusing.

(　　) ④They seldom eat locally grown food.

3. Listen again and fill in the blanks.

Cindy: Mum, have you seen ________?

Mum: I'm ________. What color are your shoes?

Cindy: Pink, like ballet shoes, they are very lovely.

Mum: Well, I'm growing ________in them.

Cindy: That's good, because I really need... ____________________?

Mum: I said I'm growing tomatoes in them. I read that a cool way to ________ something was to plant things in it. We'll soon have some nice juicy tomatoes.

Cindy: Mum, the idea of reusing is that you find something else to do with an item after you're done using it, like giving books to a library, or , or turning jam jars into storage containers.

Mum: Yeah, but ____________. You're always saying we should eat locally grown food. You don't get much more local tomatoes grown in your own shoes.

Tips

①I said I'm growing tomatoes in shoes. 我说我在鞋子里种了番茄。

②We'll soon have some nice juicy tomatoes. 我们很快就有一些多汁的番茄。

4. Complete the dialogue.

Michelle: ________________________________(你知道一些关于回收的东西吗)?

Recycling isn't just about cans and glass and newspapers.

Angela: Oh,________________________________ (回收是什么意思)?

Michelle: Well,____________________________ (回收意味着使用旧东西制作成新东西).

Angela: OK. ________________________________ (那重新使用呢)?

Michelle: The idea of reusing is that you find something else to do with an item after you're done using it, like giving books to a library, or ____ (捐赠衣服或者把果酱瓶储存到容器里).

Angela: Right, ____________________ (我们可以做得更多) more than just sort bottles, cans, clothes and paper to send to the recycling center.

5. Role play and talk with your partners with the following sentence patterns.

... can cut up... and made... into....

That's recycling!

Reusing, like giving books...

Do you want to grow... in the... ?

We can sort... to send to the recycling center.

Reading and Writing

1. Look and discuss.

Which one does not belong to organic cotton?

A　　B　　C　　D

2. Read the passage with these questions.

①Do you have any experience on clothes recycling?

②Do you have any experience on reusing?

What Is Textile Recycling?

Textile recycling is the process by which old clothing and other textiles are recovered for reuse or material recovery. It is the basis for the textile recycling industry. The necessary steps in the textile recycling process involve the donation, collection, sorting and processing of textiles, and then subsequent transportation to end users of used garments, rags or other recovered materials.

The basis for the growing textile recycling industry is, of course, the textile industry itself. The textile industry has evolved into a $1 trillion industry globally, comprising clothing, as well as furniture and mattress material, linens, draperies, cleaning materials, leisure equipment and many other items. Unlike mature industries such as metal recycling that have a long history, the growth of textile recycling is much more recent. It is worth noting, however, that a degree of textile recycling was practiced some 200 years ago in England by "rag and bone men" who collected old clothing.

The importance of recycling textiles is increasingly being recognized. Over 80 billion garments are produced worldwide annually. In 2010, about five percent of the U.S. municipal waste stream was textile scrap, totaling 13.1 million tons. The recovery rate for textiles is still only 15 percent.

In Canada, for example, over $30 billion is spent on new clothing each year, translating to approximately 1.13 billion garments, and with an average growth rate of 5.16 percent per annum. On average, Canadians discard seven kilograms (over 15 lbs.) of clothing per capita each year. Textiles constitute five percent of municipal solid waste by weight.

As such, they, along with other segments of the solid waste stream such as organics are a significant part to be addressed as we strive to move closer to a zero landfill society.

3. Decide true (T) or false (F).

( ) ①Textile recycling is the result that old clothing and other textiles are recovered for reuse or material recovery.

( ) ②The textile recycling industry has a long history.

( ) ③Over 60 billion garments are produced annually, worldwide.

( ) ④Textiles constitute five percent of municipal solid waste by weight.

4. Choose the best answers.

①How many steps are mentioned in dealing with the textiles? ________.

A. Five　　B. Four　　C. Three

②The textile recycling was practiced some ________ years ago in ________.

A. 100; America　　B. 150; France　　C. 200; England

③The recovery rate for textiles is still ________.

A. only 15%　　B. almost 10%　　C. hardly 15%

④On average, Canadians discard ______ kilograms of clothing per capita ________.

A. 6; each day　　B. 7; each year　　C. 5; each month

5. What items are referred to in the textile industry?

	Items	
The textile industry	①	⑤
	②	⑥
	③	⑦
	④	⑧

6. Answer the questions.

①What's textile recycling?

____

②Do you know some other items of recycling? Please list them.

____

③What recycling organizations do you know?

____

Further Study

1. Look and discuss.

Which pictures do not belong to Russian traditional wedding clothes?

A

B

C

D

E

F

2. Look and write.

a. felt boots　　b. shawl　　c. kokoshnik
d. scarf　　e. bast shoes　　f. sarafan
g. homespun woolen skirt　　h. Russian fur cap

☐

☐

☐

3. Find the correct words for the objects in the pictures.

Red Square vodka borscht
Lake Baikal matryoshka dolls caviar
Winter Palace Swan Lake the Kremlin

______

______

______

______

______

______

______

______

______

4. Design one of the garments with the elements of Russian culture.

5. Describe your work with a necessary introduction.

Introduction may be as follows:

- Russian elements in my mind...
- Design (colors, tones, typies...)

Sentence patterns may be as follows:

- My favorite Russian element is...
- This is a...
- My design focuses on...
- I use... because its advantage is...

My Progress Check

1. Words I have learned in this unit are:

☐ donation ☐ sort ☐ rag ☐ process
☐ export ☐ organization ☐ slogan ☐ amount
☐ citizen ☐ period ☐ item ☐ weight
☐ sandals ☐ thread ☐ enlighten ☐ quilt
☐ python ☐ coziness ☐ reuse ☐ transportation
☐ globally ☐ furniture ☐ discard ☐ landfill
☐ container ☐ strive ☐ address ☐ segment
☐ organic ☐ kilogram ☐ average ☐ approximately

All together I know ________ words.

More words I know in this unit are:

____

2. Phrases and expressions I have learned in this unit are:

☐ used clothing ☐ mosquito net ☐ recycling center
☐ waste textile ☐ plastic button ☐ soccer shirt
☐ recovered materials ☐ metal button ☐ ballet shoes
☐ clothing recycling container ☐ translucent paper ☐ jam jars

Great! Now I know ________ useful phrases and expressions.

More useful phrases and expressions I know in this unit are:

____

3. I can:

☐ identify the procedure of clothing recycling.
☐ identify clothing recycling containers.
☐ talk to a friend about recycling.
☐ talk to family members about reuse.

4. I even can:

☐ learn to know textile recycling.
☐ work together to design a garment with Russian elements.